THE WHITE BADGER

Within 25 kilometres of the centre of
London lives Snowball, a rare albino
badger.
Gary, Phil and Gordon have known
Snowball since he was only a few weeks
old. Gary was eleven and Phil fourteen
when they arrived on Gordon's
doorstep asking to be taken
badger-watching. On their very first
outing one of the boys saw an
all-white badger cub – and that is how
it all started . . .

THE
WHITE BADGER

A true story
narrated by Gary Cliffe
and written by Gordon Burness

Illustrated by Conrad Bailey

TRANSWORLD PUBLISHERS LTD

THE WHITE BADGER

A CAROUSEL BOOK 0 552 54130 3

Originally published in Great Britain
by George G. Harrap & Co Ltd

PRINTING HISTORY
Harrap edition published 1970
Carousel edition published 1971
Carousel edition reprinted 1971
Carousel edition reprinted 1978

Carousel Books are published by Transworld
Publishers Ltd.,
Century House, 61–63 Uxbridge Road,
Ealing, London W.5

Made and printed in Great Britain by
Cox & Wyman Ltd, London, Reading and Fakenham

CONTENTS

1.   Meetings   9
2.   Proof   20
3.   Snails for Breakfast   26
4.   Dogs . . . Friend and Foe   32
5.   Nights under an Artificial Moon   44
6.   Snowball the Television Star   54
7.   The Hide   62
8.   Fred the Fox   70
9.   The Iron Samaritan   76
10.   Snowball has Visitors   88
11.   A Matter of Life and Death   95

ILLUSTRATIONS

Photographs

One of Gordon's paintings showing a section through part of the labyrinth of tunnels and chambers in a typical badger set

Snowball returns, muddied, to the set after a night's foraging

Snowball, a portrait

Snowball, almost full-grown, follows his mother down the scrub bank

Fred, the fox, fully recovered

Gordon takes a photograph by remote control using a camera screwed to a tree outside

Gordon prepares for a night's filming

The old boar and Snowball's cubs, seen from the hide

A tense moment. One of Snowball's cubs actually feeds from my hand!

Badgers are welcome in forestry plantations. These swing gates keep out rabbits and foxes – but not badgers!

Snowball, the 'miniature polar bear'

Snowball's mate, her face still dirty after excavating the set

Another unusual badger! You can just see a faint stripe down its back

Our first success with the 'Iron Samaritan'

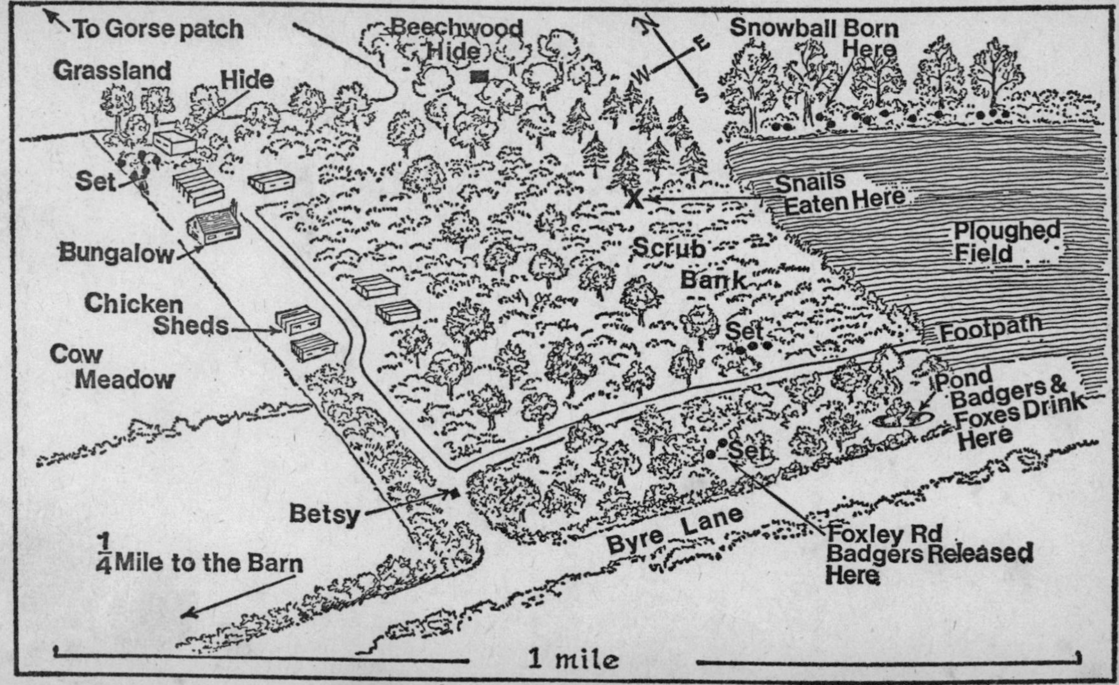

Map showing Snowball's various homes

Labels on map:

To Gorse patch

Grassland

Hide

Set

Bungalow

Chicken Sheds

Cow Meadow

Betsy

¼ Mile to the Barn

Beechwood Hide

Snowball Born Here

Ploughed Field

Snails Eaten Here

Scrub Bank

Set

Footpath

Pond Badgers & Foxes Drink Here

Set

Foxley Rd Badgers Released Here

Byre Lane

1 mile

MEETINGS

I HAD never visited a complete stranger before, and as Philip rang the bell I couldn't help wondering how we would be received – if, indeed, there was anyone at home. Then, after long seconds, the door opened, revealing a large man dressed in a sort of faded green tunic. He stood there regarding us with such an annoyed expression that I nearly retreated then and there. But not so my elder brother. Calm and collected as always, he spoke up firmly.

'Good evening. We're sorry to disturb you, but are you Mr. Burness?' The man gave a brief nod, then leaned heavily against the door, gazing past us into the rain.

'We saw your photograph of a badger in the local paper this morning, and as your address was there, we thought we'd call and ask if you would tell us how to find some. . . .' To my intense relief, at the word 'badger' the scowl had vanished from the man's face and he now began to smile.

His voice suddenly boomed, 'Badgers, eh? You'd better come in. Get some of those wet things off and tell me who you are and all about it.'

We followed the mysterious naturalist into the hall.

'I'm Philip, I'm fourteen,' explained Phil, showering us all with water as his coat came off.

'And I'm Gary,' I almost whispered. 'I'm eleven.'

'My name's Gordon and I'm older than both of you

put together. Come and sit down while I finish repairing this camera. It got rather bent last night as I was climbing down from a tree.'

The idea of this portly grown-up clambering about among branches in the dark seemed so funny that I only just managed to suppress a giggle. We sat in silence for a minute or two, then Gordon closed the camera with a snap.

'Done it!' he exclaimed. 'Now, tell me how you got interested in badgers.'

So, between us, we did. We explained to him that wild life had always fascinated us, but that there had been little opportunity to see much of it, what with school and living in a town. We had tried to make up for this during holidays and had seen field mice, voles, a fox, and quite a lot of different birds. Badgers, however, were in another class; almost magical creatures, to be found only in books. We knew something about them – that they live in underground 'sets' and come out only at night; that they eat almost anything, and are the biggest British carnivore, sometimes weighing as much as forty pounds.

'But where,' I asked our new friend, 'did you manage to photograph your badger? Surely they can't live near here, on the edge of London?'

'You'd be surprised,' was his smiling reply. 'But you seem to know quite a lot about them already. How, exactly, can I help?'

'Take us badger-watching,' I blurted out, realizing at once from Phil's frown that I had forgotten my manners. But I had burned my boats now and waited for the almost certain refusal.

Gordon spoke slowly.

'Badger-watching means mud,' he said. 'And rain, and gnats and mosquitoes, and cold winds, and long,

hard walks in the dark, and blank nights, when badgers
don't come out. Are you sure you want to go?'

'Yes, please,' we chanted together, which made him
laugh.

'All right, I'll tell you what we'll do. Last week I discovered a big badger set about five miles from here, at a
place called High Beeches. I haven't watched it yet, so
we'll have a try next Saturday. I'll pick you up at your
house at four o'clock, have a word with your parents,
and then there'll be plenty of time left for us to look
round the set area before our first watch.'

It was still raining when we left Gordon's house that
evening, but I didn't care. All I could think about was
the following Saturday. Phil became quite excited too,
which was unusual for him, and he kept rushing past me
on his bigger bike, saying that he was going to save up
for a camera and better binoculars than the ones we had,
and that we must take up Natural History *seriously*.

We were both ready by a quarter to four on Saturday.
Phil wore his leather ski-cap, a very long, old blue mac,
and plimsolls. He sat reading a bird book and looked
extremely odd. We had packed everything we thought
vital to the expedition in a haversack – sandwiches, flask
of tea, a torch and wellingtons, in case it rained – so
there seemed nothing else to do but wait for Gordon.

As the minutes slipped by I became convinced that he
wouldn't come. I had just turned gloomily away from
the window for the hundredth time when an extra-
ordinary throbbing noise, which had been getting louder
and louder, seemed to become stationary outside our
house. It stopped as I rushed back to the window and
saw a crash-helmeted figure climbing off a large and
very old motor-cycle combination. It was Gordon.

What with talking and forgetting things, it was
another half hour before we set off. Phil had let me win

the argument as to who should take first turn behind the rider, and he now peered incongruously through the sidecar window. Gordon had brought a spare crash-helmet for the outside passenger, and I felt very important in it as we roared through the town centre.

'What make of bike is it?' I shouted in his ear, and just caught the reply, which sounded like 'Betsy'. He told me later that he had said BSA, but I had rechristened her, and we called her Betsy from then on.

The journey to High Beeches took only twenty minutes, and soon we were turning off the main road on to a rough track that was signposted to a poultry farm. I nearly fell off several times as we bumped and lurched over the uneven surface, and was quite thankful when we pulled up after fifty yards or so, where a public foot-path forked away to the right from the main track. Phil stepped from his prison looking rather pale and said that after that ride he now felt capable of bringing a space capsule back to earth.

We locked the extra clothing we would need later in the sidecar and then Gordon motioned us to stand still, quietly telling us that he always did this for a minute or two on entering a wild place, to atune his ears and adjust himself to the 'atmosphere'.

None of us could have guessed on that sunny May afternoon what adventures were to befall us here in the months to come.

The footpath tunnelled its way in a straight line through several hundred yards of dense woodland, and as we walked along it, there were so many birds singing and calling that Gordon thought it worth while writing down all the different species in his notebook. I was not able to help much, but Phil recognized some of the songs and by the time we got to the end of the wood I was astounded at the length of the list: blackbird, dunnock, robin, garden-warbler, green and great spotted wood-peckers, great, blue, coal, and marsh tits and a yellow-hammer.

At the end of the wood the footpath continued across a large, ploughed field, from which direction I could hear the faint but ugly noise of heavy motor traffic. The field rose gently to our left and Gordon now pointed in that direction.

'See the big boundary hedge at the top of the field?'

We nodded, blinking in the strong light.

'Well, if we all do everything exactly right from now on, we should be watching badgers there tonight.'

He led us up the field. To our left an impenetrable bank of tangled scrub spilled through and over the re-straining fence, and as we approached the hedge I could see that it was a giant, twenty feet deep, and steeply sloping. It formed a right angle with the scrub bank and looked as if it had actually grown out of it, yet its five tall elm trees, sprouting above the lesser foliage at regular intervals, gave it an air of eerie distinction.

When we came to it the first thing I noticed was a narrow, well-trodden path, cutting through the nettles at its base. This worried me.

'Lots of people have been here. Do you think they disturb the badgers?' I pointed at the path while Gordon had a closer look.

'Good observation, Gary. But look – it passes *under-neath* those low branches, so no human could have made it. You have found the first badger path.'

We turned along the hedge, looking for more signs and soon came to a place where the solid wall of veg-etation appeared slightly thinner. Here Gordon put his finger to his lips, beckoned to us, then disappeared, soundlessly, into the green fastness without seeming to disturb a single leaf. Phil followed, pushing with some difficulty, so that I, being much smaller, was able to draw level with him before he had moved six feet. Then we both stopped in amazement. There was Gordon, clearly visible, several yards away. The hedge was hollow.

The reason for this – that not enough light penetrated the outer canopy of leaves to allow growth underneath – did not occur to us at that moment, for our attention was held by an even stranger sight. Gordon was looking up at an enormous mound of sandy earth, seemingly dumped on the bank above him. His voice echoed quietly in the natural corridor.

'This one's being used. Go up *carefully* and have a look.'

We climbed the yielding sand on all fours and, reaching the top, saw that it formed a firm, flat platform, sloping back the other way. As I craned my neck still further, the incredible truth presented itself; that the ton or more of sand we stood on had been excavated from the bank itself, and we were looking into one of the mouths of a badger set.

The hole, over a foot in diameter, bore unmistakable evidence of frequent use, its entrance belled into a polished funnel by the passage of furry bodies, its floor indented with many overlapping footprints. These prints, half as big as my hand, extended all over the platform and along the bank either side of the hole, forming the main badger highway. Lowering my head to examine them more closely, I detected a faint but exciting musky odour coming from the sand.

There were twenty-five more holes strung along the hundred yards of hedge, but after examining them all Gordon decided that only six of them were in regular use. He explained that badgers lead such hygienic lives that they often move to a different system of tunnels so as to allow fresh air to sweeten the old ones. This reminded him of another example of clean badger habits, and after searching carefully along the hedge bottom he pointed out a group of little pits scraped in the ground, filled with dung.

The mouth of a badger set

'You'll always find badger lavatories at a respectable distance from a set,' he told us. 'And, of course, they are a good indication of how many animals live there. I would guess from these pits that there are two families, one at the west end, where we came in, and one at the east.

'Now, if our latest find hasn't put you off, I think it's time for a snack, don't you?'

It was no good protesting that we had brought our own food.

'Eat it later,' Gordon suggested. 'I found quite a good café last week on the main road. It's only a couple of fields away. Come on, it's my treat.'

During the snack, which turned out to be a huge meal, we discussed the 'do's and don'ts' of successful badger watching:

Always keep down-wind.
Tread slowly and lightly when approaching or leaving a set.
Never move at all when 'in position'.
Use all available cover and do not become silhouetted.
Don't wear brightly-coloured clothing.

These were basic rules, which, as Gordon emphasized, really meant becoming an animal oneself.

It all sounded so exciting that when, at half-past seven, it was time to go, I leaped out of my chair, nearly forgetting to finish the last cake. Philip wouldn't be hurried, however, and still sat there, deep in thought, while I tugged impatiently at his sleeve.

'I've an idea,' he said. 'Would it be all right if I watch one end of the hedge while you two watch the other?'

'Of course,' replied Gordon. 'There'll be a better chance of seeing something. But if you feel nervous, give a whistle like this' – he imitated a curlew exactly – 'and we'll come and rescue you.'

Several people looked round in astonishment as the piercing call echoed through the crowded café and we left rather more hurriedly than we had come in.

The sun was dipping behind the trees when we arrived back at Betsy, so we put on our extra clothes before making our way along the familiar, but now rapidly darkening, footpath.

'Now,' said Gordon when we reached the field. 'You

volunteered, Phil, so choose your set. West on the left, east on the right.'

'West,' replied Philip bravely.

Ten minutes later we crept away to the other end of the hedge, leaving Phil safely in position with a good view of the first hole we had found and with final instructions to stay perfectly still.

The last big elm offered the best concealment at the east end, being just below the two end holes and about twelve feet away from them. But we did not stand behind it. This would have meant sticking our heads out from the side, and they might have been seen. Gordon crept up to it infinitely slowly and stood with his back against the trunk, facing the holes, motioning in the gloom for me to stand in front of him. He held his watch an inch in front of my eyes. It was five to nine. The loudest sound in the hedge now was the ticking of this watch, and as he removed it and put it in his pocket my heart took over, thumping so noticeably that I thought the badgers must hear it.

As I stood staring intently at the black entrances I could feel the steady, all-important breeze on my face, which meant that the badgers couldn't smell us – if they came out. Suddenly, for a brief moment, I thought that one of the entrances didn't look *quite* as black as the other. Yes! A chill of excitement swept over me and my heart seemed to stop as, squeaking and yapping, three baby badgers tumbled out of the hole.

Immediately they began to chase each other round the beaten platform, looking like grey, furry footballs, each with a striped, wedge-shaped head sticking out of one end and a ridiculous powder-puff tail at the other. The play was rough and boisterous with no rules and no holds barred. They bit, pushed, and jumped on each other till I feared they must surely hurt themselves.

Suddenly the smallest one got such a sharp nip that, emitting a high-pitched shriek, it leapt right off the edge of the platform and rolled, head over heels, almost to my feet.

I then did something for which I could have kicked myself but which, as Gordon said later, was forgivable in the circumstances. Half expecting the little animal to crash into me, *I moved my foot*. The cub sat up less than a yard away, peering at my legs, uncertain what to do next, while the other two stared at us, motionless, from the platform. Then, with a further shock, I became aware that the hole had filled again as an enormous adult badger as big as a bull-terrier padded silently out. This brought my cub to life, but as he scrambled back up the mound I knew we had been discovered.

The beautiful, broad head of the adult faced us, moving slowly up and down and, as if at a signal, the three cubs raced down the hole. For one long second after they had gone the big badger remained, then, after a loud snort and a thumping of paws like drums on the packed earth, the platform was empty again.

I was shivering as we left the hedge and only then thought of poor old Phil. Had he seen anything? Had he been frightened? We stopped half-way between the sets and Gordon gave two low curlew calls. For a few seconds there was no answer, then, amid the sound of breaking twigs, a distant figure, black and ragged, seemed to disengage itself from the hedge and stagger down to the field.

We ran up to him, then walked away from the set so that we could talk. I couldn't wait to hear his news.

'What happened, Phil? Did you see anything? We had a marvellous time.'

Phil seemed as composed as ever, though I thought I detected a note of excitement in his voice as he replied, 'Well, I didn't see a thing till just before you whistled, then a little badger came out and ran past me towards the next hole. It was a white one.'

Gordon stopped in his tracks.

'I think you must be mistaken, young Phil. Cubs are a lighter colour than adults, but you must have seen some markings. What about the facial stripes?'

'There were no markings,' replied Phil, a little indignantly, 'no stripes. If that wasn't a white badger cub, then I've seen a miniature polar bear!'

2

PROOF

WHAT *HAD* Phil seen? I was so anxious to know what our next move would be that I only vaguely remember the long, groping walk back through the pitch dark wood, followed by the hectic sidecar ride home. As we said good-bye I had to be content with Gordon's promise that he would come back next day for a conference. His last words were: 'Don't tell a soul about the white badger.'

The meeting took place next afternoon, round our dining-room table. Gordon had arrived in high spirits carrying an enormous book which he ceremoniously placed in the centre before he called the conference to order.

'No questions for a minute, lads,' he pleaded, as I began to fidget and point at the book.

'I think it is just possible that Philip saw an albino last night.'

He paused, and I seized my opportunity.

'Gordon, what's a bambino?'

'Wait and I'll tell you,' he laughed. 'The word is *albino*, and it means an animal which has no colour in its skin or eyes. Let me explain. I expect you know that every living thing is made up of tiny little bags called cells, so tiny that they can't be seen with the naked eye. You and I are made of millions and millions of them. Each cell has a particular job to do in whichever animal or plant it is part of. For instance, blood cells carry food and oxygen, skin cells multiply to repair injuries, and so on. In fact, they all live together in harmony as a family group, whether the group happens to be an elephant, a daisy, an oak tree, Gary Cliffe – or a badger. The reason why all these masses of cells behave themselves is that inside each one is a tiny bundle of particles, a sort of cell-brain. The particles are called genes and each one controls some part of the cell's existence, telling it where it is, whether it should move, when to repair itself, how to soak up food, what colour to be, and so on. The last one I mentioned is important to us. One gene in each cell tells it what *colour* to be.

'When a plant is fertilized or an animal mates, special cells pass from the male into the female. There are special female cells waiting for them and, if one of each happens to meet, they merge together and after a little while start to grow, dividing into a bundle of new cells which will eventually be born as a new flower, fish, human, or whatever. It will, of course, grow according to the mixture of gene-instructions from father and mother. So, if a man has dark hair and his wife fair, the hair colour of their child will depend on which gene happened to have been stronger. This means its hair might be dark, fair, or any shade in between.

'Now for albinism. Sometimes, very rarely, a gene might be missing, for Nature is not perfect. So let's see what can happen if a colour-gene is missing in one of the special cells that are used in mating in, for instance, a male badger. The cell passes into the female and is lucky enough to meet one of hers. Now, so long as her cell has the usual colour gene, the little cub, when it is formed, will be normally coloured but if, by a fantastic chance, her cell lacks a colour gene as well, the cub that grows from the two cells will have no colour-instructions in any

**Black to Brown Areas
with some Grey Hair**

Legs Black

**No Marks
on Head**

Hair completely White

**Pink
Nose**

Pink Eyes

**White
Claws**

Badger markings—the 'normal' badger and the albino

of its cells and will, itself, be colourless. An albino.

'Don't forget, by the way, that I am talking about surface colouring. The red colour of blood is controlled by a different gene altogether. That is why an albino's eyes look pink: we are seeing the millions of tiny blood vessels behind its transparent eyes.'

I think Philip understood more of this than I did and I decided to ask him to go over it again later. What I was dying to know was how rare was a white badger, and when were we going to the set again. I breathed a sigh of relief as Gordon went on.

'This is the only picture of a white badger that I've ever seen or heard of.' He opened the big book and slid it across the table.

'The book was published in the nineteen-twenties, when they used to dig badgers out of the ground and do all sorts of other rather cruel things. The photo is in black and white, of course, and if we get a colour shot I think it will be the first one ever taken.'

The little photograph showed a woman in very old-fashioned clothes sitting in a chair. On her lap, with a heavy chain round its neck, was a white badger. I didn't like looking at it very much, for it had such a miserable expression on its face, and I thought how selfish it had been to dig up a wild animal for a pet. But Gordon's next words cheered me up again.

'The next thing to do is to find out exactly what Phil saw, so, if the wind direction is all right, we'll try for a colour photograph tomorrow evening.'

Next day the usual southwesterly winds died away, leaving the air sultry and still – impossible conditions for watching at the holes. Then, as evening came on, a faint, easterly breeze took over and we remembered the run that led from the hedge into the scrub bank. We made for this spot at sunset, forcing our way through the outermost tangle of old man's beard, spindle, and chestnut saplings well down the field. Once inside the scrub bank, the going became easier, with many clear patches of luxuriant grass sprinkled with bluebells. We zigzagged up the slope towards the small fir plantation which grew level with the hedge, and soon three pairs of plimsolled feet were treading slowly and softly through the young larches to the place where we hoped to meet the badgers.

Gordon's plan was simple. The cubs we had seen two nights before were obviously adventurous, and old enough to play and forage away from the set. Instinct would direct them to do it in the safest possible place, and as the set Phil had watched was only twenty yards from the scrub bank, what better choice could they make?

Our hopes rose even further when, creeping to the edge of the plantation, we saw the well-worn path entering the trees. All the badgers had to do on reaching the end of the hedge, was to pass under the trunk of a huge fallen elm, cross a six foot clearing, then walk up a three foot slope to where we now stood.

During the next ten minutes Gordon painstakingly assembled tripod, camera, flashgun, and a long-distance air-release. Then he checked exposure settings and trained the camera on the spot where the path levelled out and entered the trees.

We were not silhouetted and the wind, such as it was, blew gently towards us along the hedge as we stood together in the gathering gloom, listening to the last two stalwart blackbirds singing defiantly at each other in the depths of the scrub below us. Then, as they too succumbed to the weariness of a busy day, the night noises took over.

A cockchafer clambered laboriously up a larch bough

six inches from my ear, making more noise than the motorcycle a quarter of a mile away across the fields. A wood mouse chiselled with machine-like frenzy at a last year's cherry stone but stopped abruptly as a tawny owl tuned up at the edge of the wood. Suddenly, there was a crash in the undergrowth over by the set, followed by a high-pitched staccato yelp. The badgers were out.

There were more crashings as the animals chased each other through the elders, this time a little nearer. Now, the indescribable sound of furry bodies padding along a leafy path. Closer and closer till, at the stricken elm, the noise stopped. A nose protruded from under the massive trunk, then withdrew. More agonizing seconds, then, with a loud sniff, a baby badger no more than a foot long wobbled into the clearing. As though on wheels it scurried up the bank, turned left, and vanished into the darkness. Almost immediately another cub appeared. This one was in such a hurry to catch up with the first that it almost overtook itself, and bounced away, emitting little squeaks of pleasure.

And then we all stopped breathing. . . .

An apparition was moving towards us from the tree. As though lit from inside, a snow-white sphere of silken fur drifted over the clearing and up the slope. I felt Gordon's hand clutch at the air-release bulb. There was a vivid flash. The little ghost sneezed in alarm, spun round and streaked back towards the set, leaving three humble badger watchers trembling with wild excitement.

3

SNAILS FOR BREAKFAST

THE FILM would take at least a week to process, but what more proof did we need? Our badger was white and rare, and somehow we had to get to know more about it.

Our first fear was that if its whereabouts became generally known, someone might try to catch it for a zoo. So we decided to draw a cloak of secrecy about the whole project and swore solemnly to each other never to tell anyone anything without having a conference first. My suggestion that we call him Snowball as a code name was accepted as a sort of joke by both Gordon and Phil, but it has since proved very useful when talking about him in cafés and other public places.

Gordon admitted, a little sadly, that he would have liked to have taken more photographs but that, as badgers never really became used to flash-bulbs going off, we would have to leave the camera at home, at least for a while, so that we could be sure that we were watching the badgers' natural behaviour.

We began to visit the hedge three or four times a week, watching from whichever position the wind-direction allowed. Sometimes we would have to stand well away from the holes and see nothing; sometimes, for no reason that we could understand, badgers didn't emerge at all during our two hour watch, and on those occasions we would creep away, wondering, and hoping they had not gone to live somewhere else.

But some nights were entrancing, and the cubs could be heard yelping impatiently below ground almost as soon as we arrived. Snowball often came out first, sometimes even before dusk, and I would nearly faint with excitement as he padded silently on to the sandy mound, lifting his pink nose to sniff carefully in every direction. Usually the other two cubs would be right behind him, and then I could hear plainly the little noses at work, testing the evening air for signs of danger.

Then would come *our* most testing moment as the cubs, as if by agreement, stood stock still and *listened*. It was usual, during these vital few seconds when the slightest sound or movement from any of us would have put an end to the evening's watch, for the dreaded whine of mosquitoes to be heard, circling invisibly above our heads. I am sure I got more than my fair share of visits from these little monsters at the most awkward moments, and I can remember several occasions when I had to resort to panicky nose wrinkling and face twitching to ward off the final landing while a great badger stood, paw raised suspiciously, a few feet away.

After this period of super-alertness the young badgers would quickly relax and begin to sniff eagerly about in the sand, turning over pebbles and sticks in babyish attempts to find morsels to eat. Occasionally one would actually find a beetle and then, in a frenzy of excited yelps, forget to eat it.

All badgers love a good scratch and these cubs were no exception. And scratching was catching! First, one would sit back on his haunches, head raised, as one hind paw disappeared in a blur of activity behind his ear. Then the other two cubs, triggered into action by the sound, would join in and soon they would all be sprawled out enjoying themselves. But when one of them attempted the 'tummy scratch', a favourite with badgers, where the animal sits upright and uses its front paws like a bear, it usually lost its balance and fell over backwards.

Although Snowball obviously had no idea that he was different, and seemed quite happy to enter into the spirit of things, it was during the cubs' periods of play that we began to notice that he always seemed to get the worst of it. The playing of young animals is a vital part of growing up, and as it usually consists of attack, defence, chasing and fleeing, it keeps them in good practice for the real thing, should they meet dangerous situations in later life.

Snowball certainly got plenty of practice in defence, for in any chase he was usually in front, and in a rough and tumble, at the bottom. But he was capable of giving as good as he got, and if one of the others sidled up to him and gave him a sharp nip, he would bounce away a foot or two, then rush back into battle. These mock fights always ended in a mad whirl, as all three chased each other's tails in a tight circle. Then, suddenly, it would all be over as they broke loose and wandered off as though nothing had happened.

Sometimes the cubs' mother emerged before we left. She was a sleek, beautiful animal, who seemed to take pleasure in her appearance, grooming herself carefully and sniffing attentively at the cubs before shuffling off to the plantation to find her breakfast. I remember that she took a special pride in Snowball, grooming him more frequently than the others, almost as if she knew him to be the product of the rare gene in her own body.

But the cubs were rapidly becoming independent, and as the weeks went by they began to depart to the feeding grounds long before she appeared. We hardly ever saw her mate, the big boar who lived in another set of tunnels nearby, and when we did, could not be absolutely certain that he was Snowball's father.

It has long been thought that badgers mate for life, but if this is so, why have there been no more white cubs in the three years since Snowball was born? It is so difficult to tell one badger from another, even in fairly good light (unless it's an albino), that we shall never be sure of the answer.

By August the cubs were almost full-grown, although they could still be distinguished from adults by their slightly more fluffy coats, which had not yet moulted away, and one August night Gordon and I decided to follow Snowball when he left the set area – an almost impossible task, we realized, but worth trying once.

That night we came specially prepared, wearing tight-fitting clothes with everything buttoned up to reduce the chance of getting caught up in thorns, and soft carpet slippers to enable us to feel things like dead twigs before putting our weight on them.

Everything started perfectly. Snowball had been coming out last for the past few nights and did so to-night. We waited for about a minute after he disappeared along his usual route, and then crept slowly after him along the main run-way leading to the plantation where we had taken his picture.

Badgers have few enemies in the wild, and because of this they don't bother to creep about silently once they are sure that Man, the one animal they *are* afraid of, is nowhere near. We were hoping to keep track of Snowball more by sound than sight. Sure enough, as we reached the plantation the sound of breaking twigs was coming from some brambles ahead. We moved on, inch by inch.

Snowball seemed to be searching for something, snuffling excitedly and causing whole patches of bramble to sway visibly as thorns dragged at his fur. Suddenly there was a loud, splintering crack, followed by the wet sound of munching. Gordon breathed one word into my ear.

'Snails.'

The White Badger ate three more of them then, tearing himself free of the brambles, padded like a grey ghost down the grassy slope towards the scrub bank, pausing once for a scratch and once to eat a few mouthfuls of grass. As he entered the bushes Gordon tiptoed forward and shone his torch into the bramble patch. There were the remains of Snowball's meal in a neat pile – not small garden snails, but the white 'Roman' species, two inches across, with shells as hard as flint. I realized then why other animals respect a badger's powerful jaws and leave him in peace.

This had cost us valuable seconds so, hurrying over to the bushes, we listened intently. Not a badger sound to be heard. I suddenly felt very daring and had an idea.

'You go round to the path in case he's gone right through,' I whispered to Gordon, 'and I'll walk down the bank and meet you at the bottom.'

There was no time for discussion. Gordon gave me the 'thumbs up' sign and then seemed to float away into the darkness, leaving me with a lot of second thoughts about my bravery.

I had explored these bushes before in daylight and, feeling certain that I could find my way, I began to follow the grassy clearings between the blackthorns. It was so dark now that I could hardly see the cruel, spiked branches in front of my eyes, and as the clearings became narrower I had the awful feeling of being trapped in a giant, natural maze.

As I stopped to collect my thoughts the night noises all around me seemed to get louder. Two shrews were doing battle in a tussock at my feet, their fierce, high-pitched squeals painful to the ears. The hoarse, unpractised 'ker-wick' call of a young tawny owl rang out above my head and, as I nearly jumped out of my skin, there was a loud thump on the ground behind me and a rabbit bolted, invisibly, away.

This was almost too much, and for a brief moment I wanted to shout 'Stop!' at the top of my voice to the hundreds of chirruping crickets all around me, for I was sure they were laughing at my predicament.

Forgetting all about Snowball now, I stumbled on down the slope, then suddenly breathed a sigh of relief. There was Gordon's torch shining in the darkness ahead. But no, it couldn't be. This light was on the ground, a brilliant, green pin-point of an eye, gazing up at me from a few feet away. In a cold sweat now, but forcing myself to step forward, I bent down and gazed back at it. It was a glow-worm, a beautiful female, clinging to a blade of grass, her body twisted so that the phosphorescent belly pointed skywards in a luminous advertisement for flying males.

The night had contained so many shocks for me so far that when, at that moment, I heard a whispered voice above me I was too numb to jump. It was Gordon, of course.

'Glow-worms are fascinating, aren't they,' he said. 'I didn't see any more of Snowball, did you? We'd better go home now. Jolly exciting, wasn't it.'

'Yes,' I answered as calmly as I could. 'Yes, jolly exciting indeed!'

4

DOGS ... FRIEND AND FOE

SEPTEMBER CAME, and with it a mild, southerly wind which lasted many days. Snowball, now seven months old, was thriving. His thick, muscular neck and scanty tail were typical of a healthy, boar badger, and his fur, no longer snow-white, had taken on a delicate peach colour, due to mineral staining from the sandy set in which he spent the long, daylight hours.

He had lately been coming out very early – dangerously so, we thought – and feeding in the lower field which contained a rich crop of clover. We could understand the attraction this must have had for him, as the damp, dark world beneath the leaves provided a home for much of his favourite foods: snails, worms, beetles, slugs, mice, rats, and the occasional hedgehog.

We always stopped when reaching the field to scan its margin for courting couples and other evening strollers, for it was part of our concern for the badgers' safety – particularly Snowball's – that we should not be seen entering the hedge.

This evening, however, it was our own safety that had been threatened in a rather dangerous incident. Bowling along at about thirty miles an hour down Church Lane, a steep hill near the end of our journey, Phil and I, both in the sidecar, had been discussing the sort of cameras we wanted to buy. Suddenly, with a loud bang, the sidecar wheel mudguard broke and jammed on to the wheel. I had a brief glimpse of Gordon wrestling with the handlebars as we bounced off the kerb and skidded sideways and then backwards down the hill, coming to rest at a fearful angle half-way up the bank on the other side of the road.

It took about ten minutes to remove the offending mudguard and persuade Betsy to go again, but we managed to coax her round in the end.

On arrival at the clover field, we gave it the usual long-distance inspection.

I don't know how Phil ever managed to see anything through our old binoculars: they had been dropped so many times (mostly by Phil), that the two eye-pieces pointed in different directions. But see things he did, and now, after slowly sweeping across the far side of the darkening field, he pronounced his verdict.

'There's a man with two dogs at the east end of the hedge.' He paused, then continued, 'They look like greyhounds.'

This was worrying. We waited several minutes, keeping out of sight but, as the man showed no signs of moving, decided to go the long way round and approach the set from the north. At last we arrived, and wasted more minutes trying to find a place to sit without being silhouetted above the holes.

Gordon was the last to roll under the barbed wire and had hardly eased himself into a sitting position when Snowball appeared on our right, oozing from the end hole like toothpaste from a tube. He wasted no time, and after a brief look round and a sniff which must have caught some tempting odours from the field below, he ambled straight down through the nettles and into the clover.

I wish now that we had stopped him, for nothing could have been easier. The slightest movement or sound from us as he stood outside the hole would have been enough to send him flying back below ground for at least another hour. But at the time we reasoned that the man and his dogs had been at the *other* end of the hedge and had surely gone by now. We did not interfere.

Once, peering through minute gaps in the hedge wall, I thought I saw Snowball's pale form about fifty yards out in the clover. I relaxed and glanced behind me, noting with some amusement that Gordon was still frozen in the 'knees-bend' position.

Suddenly, from way out in the field to our left, came a sound which turned my blood to ice – the rhythmic whispering of animal legs travelling at high speed across the field. I breathed a silent, frantic word to Gordon, who clapped a hand to his head in despair.

Greyhounds!

And we could do nothing.

The snarling began long before the dogs reached their target and as two horrible facts hammered at my brain –

that greyhounds are hunter-killers by nature, and that badgers are not afraid of dogs – I felt useless tears running down my cheeks. But as the animals met the snarls, rising to an hysterical fury, suddenly exploded into barks of astonishment and pain, and then stopped. There was now only the sound of violently disturbed vegetation, as though things were running in all directions. Something was coming towards us and I leaned forward, not daring to imagine what had happened. The nettles shook as a body passed through them and a moment later Snowball trotted unsteadily into view.

Hauling himself over the mound of sand outside his home, he paused to look over his shoulder towards the now deserted field, then slowly walked into the hole. But not before we had seen the great, dark wound on his forehead, the scar which he was to bear for the rest of his life.

Though the dogs had met their match that night Snowball had learned a valuable lesson. From then on he became warier than a fox and never again left the set before dark. Even so, by November the nights had drawn in so much that badgers were out and gone by half-past seven, and on weekdays we were often unable to reach the set in time. Thinking about this snag one day, Gordon remembered reading somewhere that many nocturnal animals cannot see red light, so we fitted discs of red cellophane to our torches and tried out the effect on our badgers.

It worked! Provided that the switch made no noise, that the light was not too strong, and that we didn't wave it about, the badgers took no notice. This was very useful, as even Snowball had been difficult to identify on the blackest of these winter nights. We were also especially pleased to be able to see that his wound had now completely healed, leaving only a thin scar.

But, what with the time of year and the weather, our visits went down to two or three a week, and seeing Snowball at all became a rare event.

The last time we saw Snowball that year, according to my rough notebook, was November 12th, and as Christmas came and went we began to worry. During a mild spell in January we mounted several long watches at all the holes we knew he used, but with no result, and when the temperatures fell again and the first snow came we did not give up.

Young badgers are usually too nervous to emerge at the first sight of snow, but from the second night onwards there is plenty of activity, and on such a night we watched the west end holes, muffled up to our ears in layers of thick clothing and with pockets bulging with flasks of hot coffee. How we would have loved to see Snowball in this matching, white world. But it was not to be.

We lasted three quarters of an hour during which time nothing moved except my teeth, which absolutely refused to stop chattering. It was so cold that by the time we agreed, unanimously, to go, our boots had actually started to freeze to the ground!

Sunday became our busiest day during this winter period. Arriving at dawn, the first job was to put food out for the birds in the beech-wood at the top of the hill where we had built an observation hide. After that we usually separated for an hour or so and carried out our own special interests.

Phil was madly keen on birds, particularly the very shy hawfinches which feed on the ground at first light. He had just joined the local natural history society, and as soon as we arrived he would rush off with binoculars and notebook to record everything he saw.

My love was small mammals, with weasels at the top

of the list. I never saw one, though, till one morning I
heard a flock of tits giving their alarm calls. I thought at
first that I was the cause of it, but noticed them diving
down in mock bravery at something in the dead leaves.
It was a weasel carrying a dead vole in its mouth. No
more than six inches long, it looked savagely beautiful as
it sat up and chattered angrily at me before diving down
a disused mole tunnel with its prey.

A weasel and its prey

Gordon usually stayed at the hide, and by the time
Phil and I got back had set up his camera for bird pic-
tures. The hide was made from pieces of canvas pinned
to stakes driven into the ground, and was big enough for
the three of us to stand inside and look out through small
holes at the trunk of an old oak tree less than two feet
away. The woodland birds had soon accepted the hide as
part of the landscape and almost queued up to get at the
dripping smeared on the bark. Our favourite visitors

were great spotted woodpeckers, who could be heard scrambling backwards down the tree long before coming into view through our spyholes. The camera had to be focused down to thirteen inches to photograph them.

In the afternoon we always paid a visit to Bill, the council patrolman who looks after the public woods across the road. We had first met him outside his little hut where, at certain times of the day, he was in the habit of brewing the most delicious 'smoky' tea on an open fire. He had been busy hanging a great, seven pound chunk of suet from a tree, surrounded by a swarm of fluttering birds. This is typical of him, for, tough ex-soldier though he is, his soul is gentle and he loves all wild life. We soon became firm friends, and when one day he told us where he hid the key to his hut, so that we could use it when he was away, we returned this feeling of trust by telling him all about Snowball and of our concern at not having seen him for three months.

Knowing Bill as I do now, I believe him when he says that sending the fateful postcard to Gordon a few days later was the saddest duty of his life. It read: *I am very sorry to have to report that a white badger was killed by a car on the by-pass on Thursday night, and has been taken to the council's animal cemetery next to the station.*

The only thing I remember about that awful journey was a feeling of numbness as I kept repeating to myself, 'There must be some mistake.'

The men at the council yards were very helpful when we explained why we were there, and two of them brought the body to us on a sack. We bent over it, hiding our feelings from the workmen.

Yes, it is a white badger – yes, there are the pink eyes – yes, it is a badger – poor Snowball – where's his scar? – I don't quite know, he was hit very hard – it should be

above his right eye – there's no sign of the greyhound
scar—

THERE'S NO SCAR!

Still pondering unanswered questions, we took the
badger back to the woods with us and buried it, knowing
that if we left it there until the skeleton became clean, we
could possibly find out from the skull how old it had
been. But by the following week foxes had started to dig
it up again, so, rather than risk losing that precious skull,
we decided on the more unpleasant but quicker method
of boiling the head till the bones were clear. This we did,
and we still have it – a beautiful, pearly-white specimen
in perfect condition. Gordon immediately took it to
someone he knows at a museum, who, after keeping us
waiting for several anxious days, gave his expert
opinion. *I have examined the teeth and inter-parietal
ridge, and come to the conclusion that this skull belongs
to a badger of more than five years of age.*

Snowball was only thirteen months old!

The old badger had been killed a mile and a half from
Snowball's home, and in the space between were two
shopping centres and two main roads, neither of them,
admittedly, enough to stop a determined badger. But we
neither knew of, nor could we find any sets in the vicinity
of the accident. Despite all these things it seems almost
certain that the two animals must have been closely re-
lated. We began to suspect that the Snowball litter may
have been the result of a brief love affair between their
mother and this white wanderer, which would prove
that badgers *don't* mate for life!

But where was Snowball?

By the time the anniversary of our finding him had
come and gone we were sure we had seen every badger
for several miles around and were forced to conclude
that he had left the district altogether.

One morning in early June we were standing around Betsy at the junction of the two footpaths when a tractor appeared in a cloud of dust, coming down from the poultry farm. We were arguing whether it made more noise than Betsy when it pulled up with a screech beside us.

Shouting at the top of his voice, the young man at the controls introduced himself as the farmer who owned the land and said he'd seen the old bike here at all kinds of odd hours during the past year and often wondered to whom it belonged.

'I don't know about badgers,' he yelled when we had explained what we did, 'but there are some holes at the top of the farm which you can look at if you like. Jump up on the tractor. I'll show you.'

There were no seats, of course, so we clung to whatever we could as the tractor accelerated away with a mighty roar, the rim of one enormous wheel spinning uncomfortably close to my nose.

Halting at the first farm buildings, the farmer leaped off to collect some eggs, suggesting that if we would like to wait, he would walk the rest of the way with us. Though hardly daring to be optimistic we realized that if there was a faint chance of finding Snowball again, this was it.

The farm stood over half a mile from the hedge set on the same side of the valley, and looking beyond the farm bungalow towards the top of the hill we could see the familiar outline of the beech-wood.

The farmer returned after a few minutes carrying a bucketful of eggs and with a large terrier of advanced years ambling at his heels. The dog was introduced to us as Spot. Spot, the farmer explained, had been with the farm when he had bought it five years ago, so he didn't know how old he was. As he patted the dog's head I noticed it was covered with curious scars, but didn't like to make any comment at the time, except to remark how much like a badger he looked with the two black eye patches on his white face. The dog was friendly enough towards us in an off-handed sort of way, but I was sure his absent-minded expression wasn't due entirely to old age. He was definitely worried about something.

We began to walk up the long path.

'Is that where you live, sir?' I asked, pointing at the modern bungalow. 'How nice to live in the woods.'

'Yes,' answered the farmer. 'My name's Bev, by the way. Yes, but I'm so busy, I don't have time to study birds and animals as you're doing!'

'What about foxes?' Gordon asked him. 'Do you get much trouble?'

'Not at all,' he chuckled. 'Our chickens are locked up at night, and on the rare occasions we have a daylight raid I usually manage to shoot the culprit the same night.'

'You must be a good shot,' I said, impressed.

'No,' replied Bev, 'I don't have to be. If a fox gets a taste for chickens I put a dead fowl outside the bedroom window, tie a cotton to it, and when we go to bed my wife ties the other end to her wrist. During the night the fox finds the bait, tugs at it, and my wife tells me she has a "bite"! Then all I have to do is reach for the twelve-bore, creep to the open window, and shoot the thief. It's impossible to miss.'

We were sure, at the time, that he must be joking, but since then Bev has had to shoot two more foxes by this method, and it works perfectly.

Behind the bungalow the ground rose steeply. And then we saw the first hole.

Phil ran forward. 'It's definitely badgers!' he called.

'Well, you have a look round,' said Bev. 'I must finish

collecting the eggs. You can do your studies here so long as you call in and let me know when you're about. Otherwise, I might shoot you by mistake! Come on, Spot.'

But the dog preferred to stay with us. He was already standing on the excavated sand outside the hole, tail wagging as he sniffed eagerly into the mouth of the

tunnel. Then, to our amazement, he crept into the set and disappeared. For a few moments there was silence, then a faint barking far away below the surface. It was not a savage bark, but had the high, anxious quality used when asking for a stick to be thrown. It didn't last long, and soon Spot reappeared backwards, shook the sand out of his coat, glanced up at us as if to say, 'That told 'em!' and trotted off towards the farm. The scars on his face were now explained. They were made by badgers!

For the next two hours we investigated the set and its surroundings. Apart from a lone hole by the gorse bushes, there were five main holes which were dug in a small chestnut and elder spinney in a corner of the farm property. The west side was bounded by a field containing a herd of cows and a bull. To the north an iron fence and grass-covered bank criss-crossed with well-worn fox and badger paths and, as we found later, the home of spotted orchids. Thirty feet to the east were two deep litter sheds full of hysterical white leghorns. And, to cap it all, thirty yards to the south was the bungalow occupied by Bev, his young wife, and their two children.

It seemed incredible that wild badgers should choose to live here, but live here they did, for the very next night we crept into a downwind position and watched a great boar emerge from the nearest hole, scratch unconcernedly, and shuffle off past the still murmuring chicken shed towards the black woods over the hill.

That was all we *saw*, but after he had gone there came the faint sound of another badger moving away from us through the chestnut spinney.

How these farm badgers slept in the daytime I just don't know. Much of the clanking and roaring from Bev's rebellious tractor must have penetrated below ground, as well as the almost continuous drone of planes from the nearby airfield; and, of course, there were Spot's social calls. And now us! We would have to be particularly careful in case our presence proved to be the last straw.

5

NIGHTS UNDER AN ARTIFICIAL MOON

A COUPLE of days later the wind turned south-west, ideal for a try at the lone hole by the gorse bushes. Gordon thought this a good opportunity for me to watch by myself for the first time, for if I lost my nerve the bungalow would be in sight! He would cross the field to the copse near the road, where he thought there were fox cubs, and try for a photo with his new electronic flash. Phil, as usual, would be birding, this time over on the scrub bank where we thought we had heard a nightjar churring a few nights before.

So it was all up to me. I walked in a wide arc round the back of the sheds so as to approach upwind, slowing as I neared the hole. At snail's pace I inched between the gorse stems and looked out on the mound of sand on which last night's intricate pattern of badger prints could be plainly seen. I sat down, leaning against the sturdiest trunk, expecting a lengthy time of peace before anything happened. Suddenly an animal leaped on to my shoulder and I reacted like a cossack dancer, almost taking the gorse bush with me, before realizing it was one of the farm cats who must have followed me. She insisted on providing an 'instant' collar which, I must admit, became quite comfortable after a while.

For half an hour I waited while the spinney grew dark and still.

All at once I felt the cat's body tense, and then she was gone. Everything was silent again.

Now there came the unmistakable sound of a badger scratching. It was outside a hole I could not see, about ten yards away.

Another outburst, this time nearer – and so was the sneezing snort a badger gives to clear its all-important nostrils. The animal seemed to be moving along a run which passed within two feet of me and ended at the open expanse of sand outside the hole I was watching. I could hear my heart beating with the same thrill that the primitive hunter must have known so well, and I stared intently with eyes now fully accustomed to the dark.

The rustle of shaggy fur brushing against vegetation drew nearer, and I prayed that the wind would hold steady. I almost felt the badger as it walked past me into the clearing ahead, and if the sky had fallen at that moment I should not have noticed. For there was Snowball! Unbelieving, I drew a sharp breath and Snowball heard it. He swung round and shot down the hole.

The time was ten fifteen, and as I was not due to meet Gordon until eleven, I continued to sit there, mentally kicking myself for causing the white badger to bolt. At ten-fifty a pale nose wavered in the dark mouth of the hole searching the air, and slowly the albino emerged, glancing briefly about him, and strode off into the night.

I had seen his battle scar!

Waiting another five minutes for Snowball to get clear, I hurried down to the fence to meet Gordon, beside myself with happiness and wondering what he would say on hearing the great news.

He was not there. A thick mist hung over the field, blotting out the outline of the copse where he had gone, so I settled down to wait, listening to the various tummy-rumblings and other muffled cow noises floating in from the white vapour.

Then three things happened all at once. The digestive peace was suddenly shattered by a loud bellowing and thumping of hooves. At the same time I remembered that Bev had said Charlie the bull was *usually* friendly, but that you could never be certain. The third thing was the appearance of Gordon's familiar figure racing out of the mist, tripod and camera-bag swinging wildly round his shoulders as he made for the safety of the fence. Charlie was about twenty yards behind him and catching up.

Unable to do anything, I stood paralysed and shut my eyes, opening them again just in time to see Gordon throw himself, with a roar of relief, at the fence – almost clearing it.

He claims to this day, and I agree with him, that it was a superb effort and that, had he not caught the toe of one boot in the top strand, that leap would probably have broken a record instead of his tripod! As it was, however, accompanied by a guitar-like twang from the fence, poor Gordon slowly cartwheeled over in mid air to land with a splintering crash at my feet. The enormous silhouette of Charlie stood five yards away on the other side of the barbed wire, the distance by which he had lost the race, and after breathing heavily for a few moments he seemed to forget all about the incident and began to graze.

Not so Gordon. 'Look at my tripod,' he hissed furiously.

'I've found Snowball,' I said.

'And what about my new flash. Ah, thank goodness! It seems to be all right.'

'I've found Snowball.'

'I thought I'd take a short cut across the field. Bev said Charlie was all right. . . . You've WHAT?' He jumped up and grabbed my shoulders.

As we walked back through the farm towards Betsy I remembered something Gordon had said last year when I had been lost in the scrub bank, and now I couldn't resist the temptation to turn the tables on him. 'Jolly exciting though, this badger watching, eh?' I asked.

'Ha, ha,' was the reply.

Observations over the next few evenings revealed that Snowball had settled in at the farm set with two normal badgers; the huge boar we had seen on the first night, and a graceful, rather timid female.

Naturally we told the whole story to Bev, who thus became the second person to share the Snowball secret. He said he hadn't known that there were badgers on the farm, and felt especially honoured that the albino had chosen to live there. If there was any way in which he could help us in our studies, we only had to ask.

So we thought, long and hard. What could be done here that couldn't be done at other sets? Of course! We rushed off to find Bev again.

'I hope you won't think we're asking too much,' said Gordon, 'but if we could use your electricity – which we'd pay for, of course – and fix a lamp outside the top shed, I think the badgers would get used to the light, and we *might* be able to take some movie film of Snowball.'

'Great idea,' said Bev. 'You can connect your wires up to the time switch in the bungalow so the light will come on and go off at the right time every night, even when you're not here.'

To complete the first part of the plan, the badgers had to be persuaded to come regularly to a prepared place outside the shed, so, clearing a small arena about eight feet away from the window, we collected all the scrap food we could find – mostly bacon, bread, and chop-bones – and carefully laid the thirty foot trail to the nearest hole. All to no avail. Within half an hour Spot, helped by two of the cats, had eaten the lot.

Now what food would suit badgers but not dogs and cats? We considered their natural diet: grass, worms, small animals, bulbs, acorns, chestnuts – Nuts! What about monkey nuts?

We bought a pound of them and returned to the farm.

'Here, Spot, have a monkey nut.' The old dog stuck his nose into the bag, thought for a moment, then sneezed heavily and trotted away. Then we tried the cats, but they wouldn't even investigate the bag's contents. So far so good. Now for the badgers.

I don't know what the top shed chickens thought that evening at roosting time as three humans slunk through the sliding doors, but judging by all the murmuring and jostling, we were certainly a subject for much conversation. They soon settled down, however, but there was a moment's panic when I found I wasn't tall enough to see out of the wire mesh window, and we had to rummage about looking for a suitable box to stand on.

We were not surprised that we did not actually see any badgers on that first evening, and felt amply rewarded, as dusk fell, at the cracking, scrunching sound of monkey nuts being eaten in the bushes to our left.

The next night was almost the same, except that we caught a glimpse of the big boar's face poking out of the leaves.

And so it went on. Little by little they got bolder until after a week both the big boar and Snowball worked their way happily down to the arena, splitting each nut neatly in two as they came to it in order to reach the kernels.

Stage one completed, we suspended a large, waterproof lamp-holder immediately over the arena about eight feet above the ground, into which we clipped a tiny, fifteen watt bulb. It took three more nights for the badgers to accept this artificial 'moon', but once they had done so we were able gradually to increase the power: twenty-five watts, forty, sixty, one hundred, one hundred and fifty, two hundred – three hundred watts!

The situation was now fantastic. The arena blazed like a film studio, the light reflecting off the shed and surrounding bushes. To our dark-accustomed eyes the effect was of a summer afternoon. Snowball, his pearly-white fur and blood-red eyes a joy to behold, basked in its radiance, lazily dissecting nuts just eight feet from our goggling faces at the window.

Because of the fascination of these goings-on we were in no hurry to risk putting an end to them by experimenting with film cameras, and for the next few weeks we were more than content just to gaze out on this wild, colourful scene.

One thing we wanted to know, for instance, was how foxes and badgers behaved towards each other, so, when foxes began to appear in our illuminated wood, we carefully noted every detail. At first they flitted across the edge of our view about ten yards away, eyes shining like headlamps as they glanced nervously at the shed. Then they began to hang about, though we never saw more than two at the same time. One old dog fox took quite a liking to the nuts and eventually attracted a litter of well-grown cubs (possibly his own) near the light by his noisy munching.

One night an adult fox came face to face with Snowball on the main run, and as they both hesitated we held our breath in suspense. But our white badger wouldn't budge an inch from the direction he was taking, and the fox, leaving it rather late, we thought, had to step smartly aside to let him pass.

After this, we witnessed many examples of the way in which foxes showed their respect for badgers, but only once did we notice any sign of friction between them. That was when Snowball became fed up with a fox who was rather dejectedly watching him eat a nut while standing less than a foot behind him. Snowball suddenly swung round with surprising speed and the fox disappeared so quickly that none of us could say which way he had gone.

Old Spot hated foxes. He had only to get scent, sight or sound of one, and he would be off after it, his hoarse bark not only wasting energy but giving his quarry plenty of time to take evasive action. Not that the fox needed any warning, for Spot had become rather lumbering in his old age; and one fox we saw dealt him a needless insult by allowing him to chase it round a small tree no fewer than six times before cantering away at quarter speed and leaving the old warrior snarling with rage.

We were longing to see Spot with the badgers, because each species obviously regarded this territory as its own. Then one night it happened while we were watching Snowball and the big boar eating their way down to the light. They were not far from the arena when we heard the dog trotting noisily past the shed door. Both badgers stopped eating and faced the point where he would appear. Spot was a forty-pound terrier and typically fearless, but the badgers did not move!

When Spot came into view scrambling up the bank into the light, he saw the badgers and froze. For about five seconds the animals glared at each other, then Spot slowly turned, slid down the bank, and we saw no more of him that evening.

This fascinating episode proved definitely that the badgers were 'in charge' during the night hours, and we could only assume that Spot's risky visits below ground were his way of claiming his authority in the daytime. They had reached a state of armed neutrality!

On another occasion one of the cats was involved in what we thought must be attempted suicide. She was crouching at the edge of the eight foot circle of light, waiting for a wood mouse who regularly raided our peanuts. Suddenly she looked away and, following her gaze, we noticed Snowball ambling towards the light, his head moving with snake-like speed as he searched for the scattered nuts.

'Watch that cat run,' Gordon whispered confidently. But he was wrong. She didn't move. It was only when the white badger snorted at her from less than two feet away that she sprang away and ran off.

Sometimes Bev gave us half a bucketful of chicken offal – messy bits and pieces left over from preparing customers' orders. Foxes just couldn't resist it, and this gave us an idea. Instead of scattering the stuff all over the wood, why not see how near we could attract a fox to the shed with it? Better still, would it be possible to photograph a fox *and* one of us in the same picture?

The arena was about two feet higher than the ground level around the shed, thus providing a low bank which might just conceal a crouching human from the eyes of a fox approaching from above. Being the smallest, I was chosen as the 'subject', and early one evening found myself squatting against the bank with my head on exactly the same level as the heap of indescribable chicken remains which Phil had tipped out carefully (and gleefully) on the edge of the bank, less than a foot from my quivering nose!

Luckily there was not long to wait. Obeying Gordon's instructions to keep perfectly still, I nevertheless couldn't help stealing a glance at the window. There was the camera, trained on me through the wire mesh and behind it the flash gun reflector and two grinning faces. Suddenly Gordon's expression became urgent and he raised a hand, pointing first at his ear and then at the sky.

I didn't need his signal, for at the same instant I realized that the air had become full of the chirruping alarm notes of swallows wheeling low over the trees. Something was on the move! Now the hysterical chattering of a blackbird joined in as it sped to another part of the wood. Then for a few seconds all was silence and in that silence, at the very threshold of my hearing, came a steady whispering, like a gentle breeze blowing through long grass, and raising my eyes I gazed into the savagely beautiful face of an adult fox.

I think we saw each other at the same time, and my spine tingled with senseless fear àt the *nearness* of this wild beast. But comedy was soon to return. As the flash went off Reynard's long legs reacted like coiled springs in an enormous backward leap and skidded into the pile of chicken innards which, flying through the air, showered over me from head to foot!

Thrilled though I was by this encounter, the next time we took this kind of picture I made sure of being *behind* the camera in the grandstand!

6

SNOWBALL THE TELEVISION STAR

THE NEXT stage in our venture was to try to film the badgers. We had just enough light, providing we used very fast film, and our subjects were willing. But cine-cameras are expensive, and we were grateful when an offer of help came from two new friends whom we had met some time before in the public part of the woods. Pete Bowman and Ricky Gandon were both expert cameramen, specializing in flower and insect photography. Not only did they borrow both eight and sixteen millimetre movie cameras for us, but they also assisted in the expense of filming.

The first camera to be tried was the almost silent, battery-operated eight millimetre version, but as Pete pressed the button Snowball bolted, scattering nuts in all directions.

But there is an answer to every problem and Gordon dreamed up an idea which, though causing a further delay, eventually did the trick. He fixed and wired up an old gramophone motor to the wall of the shed and fitted a propeller made of silver paper to its spindle. Then he put in a screw very close to the spindle so that when the motor was switched on the propeller kept hitting the head of the screw. The result was a noise just like that of a loud cine-camera! The whole contraption was wired to the time-switch which automatically switched the electricity on at dusk and off again at dawn. The badgers had begun to ignore it after four nights, and the next time we poked the little camera out between the window mesh we took fifty feet of film – probably the only albino badger 'movie' in existence.

We were never able to do it again. The next night we waited, camera at the ready, looking out at a steady downpour, and we knew what that meant: the badgers had gone in the other direction into the field to search for their greatest delicacy of all – earthworms, brought up to the surface in their thousands by the teeming rain.

This weather lasted several days, but when the ground did begin to dry and badgers still refused to visit the light it was time to investigate.

Phil volunteered for the task, and one glorious September evening he quietly left the shed and stole through the spinney that was tinted crimson by the sunset. He was gone an hour. And what a tantalizing hour for us! Almost as soon as he was out of sight the badger noises began. There were such yelps, yappings, and crashings in the undergrowth that we began to think Phil must have joined in. Apart from one brief glimpse of Snowball bounding like a gigantic ferret round a distant clump of bushes, we saw nothing of the cause of this pandemonium, and when Phil at last returned we hung on his every word.

'There was a sort of game going on when I got there – between Snowball and the little sow. And I had to be jolly careful as I got near them as they kept rushing away from the holes and several times almost crashed into me! Snowball was doing most of the chasing, but I knew it couldn't be a fight because every time they returned they groomed each other and made that purring noise. About a quarter of an hour before I left they went out of sight behind some bushes and stayed there, making yapping sounds, so I circled downwind to see what was happening. What I saw made me realize that I had been watching badger courtship, for there was Snowball mating with the little sow!'

We had tried to keep Snowball's existence a secret for as long as possible and were indeed surprised at our success so far. But now people began to talk. First it was a chemist who processed one of our colour snaps.

'I say, I hope you don't mind my asking, but where on earth did you take this picture? It's an albino badger!'

Then one of the farm customers stopped us in the lane.

'Excuse me, you're the badger watchers aren't you? I've often seen badgers crossing in front of me as I drive along Byre Lane, but the other night about midnight I distinctly saw a white one!'

From then on the news seemed to spread like wild-fire, but we were not too worried. Snowball was, after all, as safe as any wild animal could be, living virtually in a private garden, and only a select few people knew *exactly* where he could be found.

But the extent to which the news would leak would not have occurred to us in our wildest dreams – not, at least, until that evening that Gordon called on us unexpectedly.

He acted very mysteriously from the start, pacing up and down the living-room, stopping to examine the second-hand cameras which Phil and I had just bought, and then pacing again.

'What's the matter?' I asked him. 'It's not Snowball, is it? You haven't heard anything bad, have you?'

'No, no, everything's all right,' he murmured. He had stopped in front of the television set.

'What sort of a picture do you get on this?' he asked.

One of Gordon's paintings showing a section through part of the labyrinth of tunnels and chambers in a typical badger set

Snowball returns, muddied, to the set after a night's foraging

Snowball, a portrait

Snowball, almost full-grown, follows his mother
down the scrub bank

Fred, the fox, fully recovered

Gordon takes a photograph by remote control using a camera screwed to a tree outside

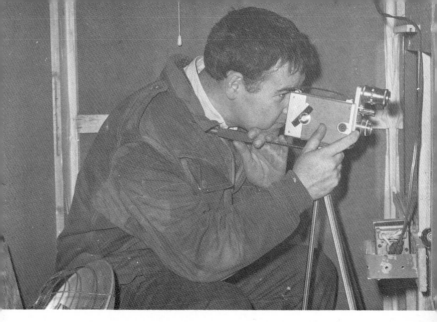

Gordon prepares for a night's filming

The old boar and Snowball's cubs, seen from the hide

A tense moment. One of Snowball's cubs actually feeds
from my hand!

Badgers are welcome in forestry plantations. These swing gates
keep out rabbits and foxes – but not badgers!

Snowball, the 'miniature polar bear'

Snowball's mate, her face still dirty after excavating the set

Another unusual badger! You can just see a faint stripe down its back

Our first success with the 'Iron Samaritan'

'Not too bad,' said Phil, 'but it goes a bit funny some-times.'

'Well,' – Gordon was grinning now – 'be gentle with it, at least till next Tuesday.'

Why, we wanted to know.

'Because *we're* going to be on it.' He chuckled.

I won't try to describe the next few minutes in which, though Phil remained his usual calm self, I am ashamed to admit I behaved like an excited baby, but out of the confusion we gradually learned what had happened.

That afternoon Gordon had had a phone call from a television producer who had said he had heard about our suburban wild life studies and also about Snowball, the white badger, and would we like to come to the studios and be interviewed? If so, a car would call for us at Gordon's house next Tuesday at two o'clock. In the meantime, if we would send some of our photos they might be able to show them on the programme as well!

'I took a chance,' said Gordon, 'that you would want to go, so I accepted. You'll have to get leave from school, of course, and Mum says you can have lunch at our house. Best bib and tucker, remember, and whatever questions are asked at the studio, don't, whatever you do, say exactly where Snowball lives, because you'll be telling about ten million people!'

At last the great day came, and by two o'clock we were all walking about bumping into each other, making jokes and pretending not to be nervous. Then I remember calling out, in a voice which I hardly recog-nized as my own:

'A gigantic car's just pulled up outside and a chauf-feur's getting out!'

We were to travel in style!

A few minutes later, settling back in the limousine's

bed-like back seat, we departed to a rousing cheer from at least a dozen of Gordon's neighbours, a send-off which, if a little embarrassing to us, certainly delighted our chauffeur!

It was this jovial Irishman, who insisted on being called Paddy, who put us at our ease during the long journey with a host of funny stories about some of the well-known celebrities whom he had ferried to and from the studios in his time. While still several miles from our destination, however, we began to catch glimpses of the imposing transmitter mast reaching into the sky, and each time my stomach turned upside down.

As we approached the television building I thought how strange it was that my idea of heaven, shared by both Gordon and Phil, was to stand in the pre-dawn silence of an English wood, heart thumping at the approach of some small animal. Yet here was this great towering mast, surging with enough power to kill a thousand badgers, and our hearts were thumping in much the same way!

We had been held up several times by dense traffic, so that when Paddy brought the car screeching to a halt outside some huge doors marked RECEPTION, he leaped out and had the doors open in a flash.

'We're a bit late,' he said. 'Jolly good luck to you, and I'll be waiting for you when you come out.'

Giving our names to a uniformed doorman, we were shown inside to the plush luxury of the reception room, but had scarcely sat down when a worried-looking man hurried over.

'Mr. Burness and Masters Cliffe? Would you follow me, please. I'm afraid we're rather late.'

The immense building seemed to consist entirely of corridors, but finally the man darted through a door and we found ourselves in the most complicated room I've ever seen. There were cameras, monitor screens, microphones, men with space-gear on their heads and ankle-deep in wires and cables. Several cameras were pointing threateningly at a small arena at one end of the studio so brilliantly lit that I suddenly yearned for our distant woodlands. It was then that I realized with a shock that we were about to undergo exactly the same kind of experience as our badgers. Now it was *our* turn to occupy the pool of light!

A voice came over the loudspeaker. 'Gordon, would you mind sitting over there with Philip and Gary on your left?'

I tiptoed forward through the cables, feeling like an early Christian being thrown to the lions.

'Please try not to knock your microphones over,' boomed the invisible voice. 'If you will look behind you, Gordon, you will see that we've projected one of your paintings for a background.'

Glancing over my shoulder, I did manage to recognize the immense portrait of a badger. It had taken Gordon twenty hours to create that little badger with its bundle of bedding, using the finest of brushes. Now the animal loomed above us, an eight-foot monster, clutching what looked like a cartload of straw to its breast!

Dragging my eyes away I cheered up a little at the appearance of a beautiful young woman who had emerged from the shadows.

'They're going to record the programme on tape,' she told us. 'Shall we run over it once to get the timing right?'

The first thing to appear on the monitors during the actual recording was our best shot of Snowball emerging from a hole. I remember that much, but I don't know how we got through the next half-hour. I suppose all that happened is locked away at the back of our minds, but I can only hazily recall Gordon, paler than I have ever seen him and looking as though his tie was strangling him, talking his way through the film sequences.

Then the young woman turned to Phil and I, asking us alternate questions about how the adventure started, what had happened so far, and what plans we had for the future.

Phil, of course, answered perfectly, but I remember making a couple of silly grammatical mistakes which caused gusts of laughter from the cameramen.

At last, the invisible voice boomed again. 'Fine, fine, perfect. Just right for timing, too. Thank you very much.'

But I don't think Gordon heard it. The heat from the powerful arc lamps had been almost too much for him, and he had subsided into a bewildered, perspiring heap in his chair!

The tension gone, we began to take a livelier interest in the busy scene, enjoying the faintly weird experience of being introduced to familiar television faces in person, yet having, of course, to greet them politely as perfect strangers.

The programme editor had a long talk with us, asking to be kept informed of further developments in the Snowball story. 'And if you ever write a book about him,' he said as we shook hands, 'be sure to let us know.'

Next we had tea and biscuits with our pretty interviewer.

'You're welcome to stay as long as you like, as far as we're concerned,' she said afterwards, 'but don't forget "your" programme will be broadcast in less than an hour and a half, and you'll probably want to see it at home.'

We had forgotten all about Paddy!

Bidding everyone good-bye, we rushed out into the labyrinth of corridors, just avoiding a nasty collision with my favourite newsreader, and somehow found the way out into the courtyard.

Paddy was asleep – who could blame him! Nevertheless, it took two piercingly whistled choruses of *Mother Macree*, delivered by Gordon two inches from his window, to wake him.

'Paddy,' he pleaded, '*please* step on it. The programme's due to go out at six fifteen and it's already nearly five o'clock!'

'What?' asked Paddy incredulously. 'You'll never do it! Have you forgotten the traffic at this time of night?'

But he tried. He ghosted us through block after block of densely packed vehicles all heading, as we were, for the suburbs. But hold-ups were inevitable, and time was beating us. The critical point was reached with six miles to go and five minutes to the start of the programme. There wasn't a hope!

But then Paddy's eyes gleamed with an idea.

'A café's the thing,' he shouted. 'I know one along this road with television!'

His knowledge of human highways, as good as, if not better than ours of badger runs, saved the day.

As the screen lit up in the obliging proprietor's back room, Gordon raised his tenth cup of tea that day and proposed a toast: 'To the wilds of outer London.'

I looked out of the window at the block of flats which seemed to occupy the entire sky as Phil's voice summed up all our thoughts. 'And good luck to Snowball.'

Our wild, white badger appeared on the screen.

7

THE HIDE

AFTER SNOWBALL mated with the little sow, he never again followed the nut trail to the light – not, at least, while we were there. *She* had always been too timid to approach the shed, anyway, and perhaps Snowball felt it his duty as a good husband to accompany her on her dark wanderings. Soon, however, our monkey nuts lost their attraction altogether, even for the old boar; for as autumn advanced, acorns, chestnuts and beechmast began raining down from the trees, supplying vital, fat-producing food for all woodland creatures as they prepared for the lean months ahead.

But one chill November night we saw a sight which gladdened our hearts: Snowball shuffling backwards with a great armful of grass towards the main hole. Taking in bedding meant that he was content with his life at the farm and intended to stay with his mate for the winter. This, in turn, meant that as all badger cubs are born in February, their cubs would be born here!

Things were not going too well in the chicken shed. *We* were happy enough, of course, protected from the weather and still seeing the occasional badger, fox and wood mouse, but we had been vaguely aware for some time that the chickens, nervous creatures at the best of times, were showing signs of lack of sleep. When one night I accidentally knocked one off its perch, it leaped into the air causing the whole flock to erupt in a sort of communal nightmare, and we realized we were not being very fair either to Bev or the hens.

Bev was very nice about it when we told him, but said that he *had* noticed that egg-production had been going down in that shed lately, and why didn't we build a proper observation hide nearer the set?

What a marvellous idea! Gordon, being an engineer, immediately began drawing up plans for the new hide which, he said, must be scent-proof if it was to be near the set. It would have to be at least six feet high, six feet wide and, seven feet long to accommodate seats and a table, and have a full length, sliding glass window. The 'air-conditioning' would consist of a tall chimney with a suction fan mounted at the bottom, so that all scent would be sucked out and blown away high up over the badgers' heads.

At this point our good friends Pete Bowman and Ricky Gandon came on the scene again and, taking a great interest in the new project, offered to do all they could to help its success.

A few days after we had drawn up sizes and shapes, Pete arrived in a van with all the sides, roof, and window panels ready to be assembled. This was very important, of course, as it would avoid frightening the badgers too much by unnecessary hammering. Now there followed a great deal of discussion as to where to put it, but, though strongly tempted to build slap in the middle of the occupied part of the set, we had to bear in mind Snowball's wary disposition. So we finally decided on a bare patch of ground ten yards from the nearest used hole, but still in full sight of it. The most important reason for this final choice, however, was the possibility that we might attract the animals to two disused holes on the edge of the bare patch, not more than twelve feet from the planned position of the hide window.

I won't describe the erecting of that hide, except to say that it took four Sundays of frustration, hysteria and minor injuries consisting of pinched fingers, hammered thumbs and grazed knuckles. There was also what I like to call the highlight of the project when, in the early stages, the whole construction came crashing down on Gordon's head causing him to swallow an enormous boiled sweet that I had just given him!

Eventually all was completed, and as Ricky Gandon pushed in one of the electrical plugs in the elaborate control panel, the suction fan started up with a gentle hum. We were scent-proofed!

By the middle of January our little house was fully equipped. There was a cupboard full of pots and pans, coffee, sugar and dried milk; a shelf holding an electric kettle and two electric fires; and along the back wall a bench big enough to sleep one person when taking turns to watch all night. On the front wall was mounted the very important 'clicker' motor, wired, by a very long cable this time, to the bungalow, as were the two power-ful lamps outside, one of which pointed down at the disused holes and the other at the occupied part of the main set.

It was no use pretending that we hadn't caused any disturbance during this latest venture, and I suppose we were lucky the badgers didn't desert the set altogether. As it was, a week of intensive watching passed before first the big boar and then the little sow began to approach the hide. Within the next two months we had taken movie film of both of them, but Snowball absolutely refused to have anything to do with it! All we ever saw of him was his ghostly figure creeping in and out of the bushes at the very limit of the light beam. But to compensate for this there was a change in the little sow's behaviour, which indicated that she had had her cubs!

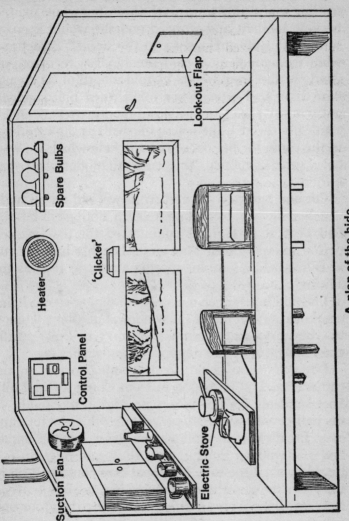

Suction Fan

Control Panel

Heater

Spare Bulbs

'Clicker'

Electric Stove

Look-out Flap

A plan of the hide

Now, instead of emerging from the occupied hole and trotting confidently towards the hide, her whole manner became hesitant and worried and she would keep returning to sniff anxiously into the tunnel. Sometimes, while eating nuts quite contentedly a few feet from the window, she would stop in mid-mouthful as though something had crossed her mind, then bolt suddenly away at full speed, her generous hindquarters bouncing comically through the young bluebells as she sped back to the cause for her concern – the hole which we were now sure contained the cubs that Snowball had fathered.

On one Saturday in mid-April we had an all-night watch and when, by half past eleven, nothing of interest had been seen, Gordon volunteered for the first rest period and was soon flat out on the bench, a half-full cup of coffee beside him, his snoring almost as loud as the 'clicker'.

Phil and I, seated on our sawn-off dining-room chairs, continued to peer out of the window, but after a while I, too, began to doze, despite the worry of Phil's steady champing at our limited supply of sandwiches.

I remember my chin slowly coming to rest on the camera fixed on its tripod in front of me, when Phil's knee suddenly crashed into mine, and in the middle of my half-conscious grab to prevent the whole lot toppling over, I realized with a jolt why he had woken me. Snowball was walking slowly across the beam of light, followed by the little sow. Maternal care could be seen in her every step, for there, bouncing in single file between them like balls of grey knitting wool, were their four tiny cubs! I turned to wake Gordon, but he was already kneeling behind us, sharing what was probably the most dramatically beautiful scene that we have ever witnessed.

Although all four cubs were normally coloured, it was thrilling to know that in each tiny body lay the latent power to produce further white badgers in the years to come.

I had heard that it was possible to feed wild badger cubs by hand but, though dying to try it with these babies, I had to plead with Gordon and Phil for some time before they agreed. I can see now that their objections were reasonable – that if these or any wild animals are tamed to any extent they will tend to trust *all* humans, including those with guns! But I was so certain that no harm would befall them on the farm that we decided to attempt this new project for a short while and then let them revert to their natural wild state again.

As the operation would take a little time to produce results, we started immediately and bought a large jar of honey which we diluted with water so that it filled two quart bottles. Then, half filling a small enamel bowl with bits of broken bread, we poured enough of the sugary liquid over it to make a sort of mash. The bowl was then placed in a dark, secluded place, out of the light, near one of the used holes and, most important, at the foot of a thick, leafy sapling.

The next night it was still there.

But on the third night the bowl was licked clean, so we filled it again and left it as before.

On the following night the bowl was gone! (We found it months later, dug out with some old bedding from the main tunnel.)

Bev's wife, Dot, supplied the next bowl and this time we drilled a hole in its rim and wired it to the sapling.

Now the food was being eaten regularly we began to watch from a discreet distance to see exactly what happened. The first time we watched it was all too easy. We had barely filled the bowl and retired into the nearest cover when a cub walked out, shuffled straight up to it, and devoured all our carefully prepared mash in about thirty seconds!

Needless to say, next evening I was waiting for him, sitting on the ground hugging the sapling and trying to look as though I had been there for years. I was determined that if anything went wrong it wouldn't be my fault, if only because Gordon and Phil were watching the whole thing through binoculars!

At exactly the same time as the previous evening a little striped face appeared in the entrance and a cub (it might not have been the same one) trotted confidently towards me. It didn't suspect anything was wrong until it was about a yard from the bowl, then it froze for several seconds as though turned to stone. Next, with a minute sneeze of alarm, it jumped clean off the ground, landing a foot farther away but still looking longingly at the food.

Now it had a sort of mental wrestling match with itself, its questing nose trying to approach the bowl while its back legs were trying to run away! The nose won, and in a few minutes the delightful creature was lapping noisily at the sweet mixture a few inches from my outstretched hand.

Luckily, the usual south-west wind held true for the whole fortnight that we fed the cubs, for otherwise I don't think we would have had so much success despite the helpful, covering scent of the food.

The next night the venturesome cub approached even more confidently, unaware that two changes had been made in the arrangements: Gordon was lying with his flash camera ten feet downwind, and my hand was inside the bowl, submerged in the food!

As the cold little snout snuffled between my fingers at the bottom of the bowl I raised my hand infinitely slowly and offered the last bedraggled bit of crust. The cub took it delicately in its jaws, gulped it down, then without any warning came back for more – my thumb! I am sure that the tiny teeth would have done no damage, but as Gordon chose that instant to fire the camera we never found out. The cub darted away, stopped, looked back with as much curiosity as surprise, then waddled back down the hole.

After this we took turns at the honoured position at the bowl, Phil being especially lucky one night in having three cubs queueing up to be fed. But the funniest and most amazing incident of all was during Gordon's first turn, when the huge boar suddenly galloped out from behind some bushes and made straight for the bowl. We will never know whether he intended eating from it because, understandably, Gordon withdrew his hand with such speed that the whole mass of soggy bread was still flying through the air after the astonished badger had reversed back into the bushes. We all know that the bite of an adult badger can crush a man's hand, but I sometimes wish that Gordon had waited just a *little* longer . . .!

FRED THE FOX

ON NIGHTS when homework, visiting relations or other duties prevented us from spending a full evening in the hide we always tried to visit it, if only for a few minutes, to throw out the usual ration of three handfuls of monkey nuts which the cubs were beginning to eat. It was on my way back one evening in pouring rain about a week after the hand-feeding adventure that I met the fox.

He was the smallest, wettest and most unhappy fox cub that I have ever seen, and hurrying past the bungalow I would surely have trodden on him had he not greeted me with desperate squeaks from deep in the long, rain-soaked grass. As I stopped he crawled to my feet and began gnawing the toes of my boots in his hunger, and when I picked up his cold, thin body I though that he must be dying.

There was only one thing to do, so I ran to the bungalow and presented the surprised chicken-farmer with this pitiful example of his worst enemy. But we were both welcome. While Bev dried him with a cloth his wife, Dot, prepared a saucer of warm milk. Then, placing both cub and saucer on the floor in front of the stove, we stood back to see what would happen. But we all forgot that Puffball, the little she-cat, had new-born kittens in a box on the other side of the kitchen, and before the cub had taken two tottering steps towards the milk the spitting puss was already on the attack, half-way across the room. The situation was saved in the nick of time by Dot neatly catching Puffball in mid-air, and when cat and kittens were safely out of the room the little fox who, despite his weakness, had prepared to defend himself, squatting back on his haunches with flattened ears and open mouth, now advanced once more on the saucer. He was on his third refill when Gordon walked in. I was supposed to have met him an hour ago!

After examining the cub Gordon said he was certain that it had either been abandoned by its parents or had got lost, and that as we obviously couldn't leave it at the farm we would have to accept responsibility for it, take it home, and care for it until it could look after itself.

Before we got to the motor-bike I had persuaded Gordon that Phil and I could take on the task and would love to do it, so, cuddling the now warm and sleepy cub inside my jacket and feeling so overjoyed that he was mine, I christened him with the first name that came to me – Fred.

Fred didn't receive the same warmth of welcome from my parents as he did at the farm, but he was allowed to spend the night in the kitchen on the understanding that we would provide a permanent home for him in the garden next day. Before we went to bed we left him curled up asleep in a box half-full of crumpled news-paper, with a bowl of bread and milk within easy reach and, as an afterthought, two chop bones under the sink.

The next morning I awoke to the sound of my father's voice.

'Come on down, you two! I think this wild animal's showing signs of recovery.'

Hardly knowing what to expect I ran downstairs.

Fred was definitely feeling better. He had overturned his box, ripped the newspaper to shreds, eaten his bread and milk, and was now trying to hide the second chop bone behind a shopping-bag in the corner. As we watched, he lost his temper with the bag and fought with it, screaming, cat-like in his fury.

We were forced to agree with Dad that the sooner we built an outdoor home for him the better!

An old rabbit hutch, suitably strengthened, served as his sleeping quarters for the next two nights while we arranged for some heavy wire mesh to be delivered. From this we made a six foot long, four foot wide, three foot high compound, and attached it to the hutch at the end of the garden.

Gordon came to help in the difficult business of bending the mesh, and soon Fred was inside inspecting every nook and cranny and, as ever, looking for things to eat.

From then on we fed him mainly on chicken heads obtained free from our local butcher and gradually accustomed him to drinking water rather than milk, until his tummy grew round and firm.

We never forgot, however, that Fred was a wild beast and that as such his behaviour was worth studying. The most interesting thing of all was his temper! Although normally playful and friendly with us, directly he saw or smelled anything resembling food he became a raving maniac; eyes blazing, ears back and teeth snapping at everything within reach. We took this to be a good sign and were pleased, because it showed the natural killer instinct that he would need when we released him.

But this was not enough. Fox cubs learn to catch their food by competition with other cubs and copying the hunting tactics of their parents. Fred never had the benefit of this experience, so it was up to us to provide it!

We began taking him with us to the woods and letting him roam around more or less as he pleased, on the end of a long rope lead.

The technique of teaching him to kill was simple enough; we just tied a piece of food – sometimes a dead bird or mouse found by the roadside – to a long string and ran in front of him, dragging and jerking it through bushes and over tree stumps, trying to make the bait look as 'alive' as possible.

What with the lead and the string we often got into some frightful tangles which sometimes took several minutes to sort out, and while this was going on Fred would curl up, yawn with boredom, and then drop off to sleep within seconds.

But he soon improved in the speed and judgment of his pouncing and when one afternoon he ignored the bait and, with a lightning leap to one side, caught and killed a fleeing bank vole, we knew his stay with us would shortly have to end.

Barely nine weeks after I had found him he made his last journey to the woods, and as he sat on my lap in the sidecar I realized how much I would miss him. But now he was a real fox. The snub, puppy face with boot-button nose had sharpened, the rat-like tail, which used to wag madly whenever I approached, had become a respectable 'brush', and his once grey, powder-puff fur had thickened and taken on the colour of autumn leaves.

We went for our usual walk with Fred, as always, tugging impatiently at the lead, and on a grassy patch at the edge of the beech-wood we removed his collar.

At first he didn't realize he was free, and sat there waiting for his 'hunting' practice. We threw six chicken heads as hard as we could so that they scattered into the brambles, and then watched him search them out, eating the first two and burying the others in the loose soil as he came to them.

As he hid his fifth chicken head deep in the undergrowth, his white-tipped tail waving gently and happily above the brambles, we coiled up his lead for the last time and left him.

About a month later I was sitting at the top of our observation ladder at the edge of the farm set looking out hopefully, as usual, for Snowball, who seemed to have disappeared again. Just before dusk I heard the unmistakable swishing trot of an approaching fox. It was coming from the field behind me and I didn't see it until it passed almost underneath the ladder. He was a beautiful male in full summer coat, and there was something about him. . . .

The sound of a human voice, especially at close range, is probably the most frightening thing of all to a fox's ears, but on this occasion I just couldn't resist the temptation. Bracing myself for the certainty of his streaking away in panic, I spoke in my usual voice.

'Fred.'

The fox hesitated, ears back, then continued – but more slowly than before.

I called again. 'Fred.'

This time he stopped and looked back over his shoulder. Was that recognition in his amber eyes?

'Good-bye Fred,' I whispered as he turned his graceful body once more and disappeared into the gathering dusk.

9

THE IRON SAMARITAN

WE FIRST heard about the badgers of Foxley Road from John French, a badger-watching friend of Gordon's from the days before he had met us.

John lived about fifteen miles away from our woods in an even more built-up area than ours, but nevertheless he had twelve occupied badger sets in his 'care' – which meant he watched them as often as possible and kept them free from dumped litter and other annoyances. He had been keeping regularly in touch with Gordon all this time and now, in a recent phone call, had asked for our help in what he described as a 'dire emergency'. Some of 'his' badgers were in danger.

On the phone it had been arranged for John to meet us in his car at the farm on the following Sunday, when we could show him the sets in Snowball's domain and then go back with him to Foxley Road, the scene of the disaster. Gordon lost no time in explaining to us the details he had received from John so far.

Foxley Road is a small lane running between two rows of detached houses. The gardens are large and on different levels. Consequently, on one side of the road the ground sloped to the lower level. In this bank badgers owned a long-established set. Apart from its being close to John's home, his special interest lay in the fact that the main set entrance opened out just two yards from a lamppost at the edge of the road!

The trouble started, apparently, in a small way. Someone threw on to the set some large flints that were probably discarded from a garden. Soon bottles, lawnmowings and bits of unwanted household machinery began to appear, but by reading the signs – dung pits, bedding, used runs, etc. – John knew that the badgers intended to stay on in spite of the accumulating rubbish. Two entrances were still clear, so he decided not to risk attracting attention to the badgers' presence by clearing anything away at that time, and left someone's latest contribution – a spin drier – lying where it had been thrown, across the bottom hole.

The development of a rubbish dump seems to be an automatic process, and it soon became evident that enthusiastic gardeners and spring cleaners were not the only people responsible. The quantity and variety of junk grew apace, and on one miserable morning a lorry driver, no doubt anxious to shorten his journey, deposited his entire load of broken tiles on the set and buried the badgers alive. This time John thought he had better help a little, so he spent the next two hours uncovering the main entrance by removing several hundred tiles, a job made more difficult by his having to keep a look-out for local householders and cars passing along the lane. John knew from experience that it is not safe to assume that everyone likes badgers, and so every time someone came along he stopped what he was doing and pretended to look like a bird-watcher.

Next day he was amazed to see that not only had the badgers completed the job of clearing the main hole to their own satisfaction, but that they had opened up *another* through the three foot layer of roof tiles farther down the slope!

Never were there such obstinate badgers! The only thing that worried John now was that the hole had become very noticeable. So he kept an extra close watch, and a few days later he was infuriated but not really surprised to find that those harmless badgers had a human enemy.

Whether this person had been disturbed by the animals at night as they tried to restore their ravaged home, or whether he was just annoyed by the sight of the holes, we shall never know. John could only stare, aghast, at what had been done. Two loads of cement weighing several hundredweight had been poured into both holes and left to set.

It was time for action to save the badgers. But what could he do? He knew that even now the badgers were in no immediate danger because they could easily dig round the cement (which they did that same night), but a human enemy hates to be defeated and John knew only too well that the next stage would almost certainly be poison or that terrible anti-vermin weapon – cyanide gas! It was time to call in help. And that was when he phoned Gordon.

We all went to examine the set and its surroundings, and on the way there John told us of his plan to save the badgers.

Gordon thought the idea sounded a bit drastic at first, but when, on arriving at the stricken set, we saw that prints in the freshly dug earth round the cement blocks were showing quite plainly that it was still in use, we agreed to take part in his plan.

We would trap the badgers alive and unharmed, and transport them to an empty set in the safety of our woods fifteen miles away.

Back at John's house, we sat down to design the trap, and after much argument and discussion and covering many sheets of paper with sketches and sizes, we came up with what we thought would do the job.

The trap would be four feet long, and two feet wide and high, with a sliding door at one end which, when triggered by a sensitive plate hanging at the other end, would drop down and close the trap.

Gordon had the main frame and the delicate triggering parts made at the factory where he worked, but when John went to collect it they couldn't fit it in the car! Eventually, to the amusement of a crowd of Gordon's colleagues outside the factory, they managed to get it three-quarters in on the back seat, with Gordon sitting bent up inside the frame and holding the car door as near shut as it would go. When they arrived back from this ten mile journey he had to be helped out. He says his back has never been the same since!

To cover the trap and its sliding door we used the strong wire mesh from Fred's old compound, but it was more difficult than we had thought to cut, fold and fix this to the heavy iron frame. By the time we had finished, Gordon had lost his temper twice and John had lost one finger nail! Finally, after filing all sharp edges and corners away, we were ready for a test. The door was pulled up on its runners until it rested on the end of the supporting rod, which ran along the top of the trap. Then I was allowed to poke a stick through the side, and as it pressed against the metal plate hanging from the rod's other end, causing it to move a fraction of an inch, the door fell with a clang. It worked!

For the next two days, Thursday and Friday, the heavy construction was left to 'air' on Gordon's lawn to rid it of human and factory smells, for on the Friday night we were going to forget our beds and put the 'Iron Samaritan' to work.

Late on Friday evening Gordon carefully greased its moving parts with lard, and with John's help (it was as much as they could do) loaded it on John's new roof rack. Then, after calling in to collect Phil and me, and after re-checking the rope lashings securing the trap, the car set off towards the unknown risks of what we now look back on as 'Operation Night Rescue'.

We arrived, as planned, after dark, and drove a short way into the lane where we turned off the lights and waited a few minutes in case we had attracted any unwanted attention. All seemed quiet so, checking our watches – it was nearly ten – we got out, untied the lashings and unloaded the trap. Then, each taking a corner we walked in double file along the grass verge, observing, of course, the golden rule of absolute silence. Just as the ache in my arm became almost unbearable, for I was sure I had the heaviest corner, the lighted lamppost came into view through the trees, and after a few more agonizing yards we lowered the trap gently to the ground.

John now crossed the lane to take up his position by one of the garden gates where he could see the set from a slightly different angle, and we all settled down to wait.

The scene looked desolate through binoculars. Two moths were whirling round and round the light, and after a while, despite my scarf and overcoat, the August night began to feel cold. I took from my pocket the watch Dad had lent me and held the fading luminous dial close to my eye.

A quarter past one – A.M.!

Across the lane a torch flashed red, and at the same time Phil's elbow pressed into my back. I raised the glasses. A grey hump filled the hole mouth, and a badger crept with fox-like stealth on to the rubbish. Then the nocturnal silence was broken by a deafening avalanche of tiles as they gave way under the poor animal's weight. There was a flurry of paws as Brock struggled to regain his balance, then he crash-dived back into the hole!

Silence fell again ... and continued. Our chances didn't look too good, but as nobody else seemed to be losing patience I shifted my weight on to the other foot and let my mind drift into that pleasant daze which is something like sleep, yet leaves the senses wide open to detect any change in the situation.

At 3.10 A.M. it was my turn to nudge Phil, whose breathing had now taken on a suspicious regularity. Two badgers had emerged from the concrete, their striped faces standing out in the lamplight. Without hesitating, they now skirted the tiles and crossed the lane in front of us with that typical shuffling trot that, to-night, sounded like a couple of lawn-mowers.

The animals made for the only well-worn run which we had been able to find. Starting from the lane, it led through the untended shrubbery of a large garden whose house loomed unlit in the darkness. As the shaggy stern of the second badger entered the black bushes we strained our ears to follow their progress. The snuffling and brushing noises, with the occasional snap of a twig, gradually receded into the distance, and I found myself comparing the disturbed lives of these animals with those of badgers in our own area. Even on these short summer nights the farm badgers were able to emerge around nine o'clock and enjoy life above ground for seven hours; but this pair of frightened creatures would have to return within two hours! They must be slowly starving.

A low curlew whistle to John brought him tiptoeing across.

'We'll set the trap in the nettles at the edge of the tiles,' he whispered. 'They should come back on the same run.'

T–E

But it's no use making just one plan when working with wild animals, and I suddenly realized that we had forgotten something.

'How can you catch two badgers with one trap?' I asked.

The fact was, that deep in our heart of hearts we had not *really* expected to catch anything! But John had an idea, just in case.

'There's an old sack in the boot of the car,' he said. 'If we catch one, we might be able to transfer it to the sack and then reset the trap. Here are the keys, Phil. Nip back and get it, will you? And if you see a torch flash on your way back, don't come any further; it'll mean there's a badger about.'

As Phil disappeared, the rest of us hauled the trap the remaining twenty yards to the set and positioned it in line with the run, its open end facing the garden from which we hoped the badgers would return.

Now Gordon walked farther up the lane on the opposite side and chose a tree to lean against, just past the spot where the badgers had entered the garden. His job was to hasten them towards the trap.

As John and I returned to our old positions, the sky was paling rapidly, and a robin began to sing, at first uncertainly, and then with its usual sad assurance.

The car boot clicked shut three hundred yards away. If I could hear it, what must such a sound be like to a fox or badger?

At this point I glanced behind me, and there was a fox sitting in the middle of the lane, watching the whole proceedings! Seeing me turn, it raced away, fairy-light on its feet, into one of the huge gardens, and as it ran its flight was marked by alarm calls: first, a dunnock's anxious piping, then a blackbird's fevered chatter. The last call I remember hearing was a jay's. As the rasping note died away a large twig broke in the garden to my left. This was it.

The timing of the next move had to be exactly right. If we mismanaged this part, the badger would crash off anywhere and be lost.

Our quarry paused unseen at the edge of the garden, and I knew that Gordon and John were aware that the critical moment had arrived.

For an instant the badger's head stuck out from the undergrowth, then he was out on the lane, trotting purposefully towards the trap.

Come on, Gordon!

Gordon's noisy step forward was perfectly timed, and in a spurt of gravel the badger accelerated straight towards the steel framework between him and the safety of the set.

Yet even in a state of panic a badger's reactions are lightning-fast, and as he entered the open door he slithered to a halt. He was turning – and the door was still open!

But, in spinning round, his shaggy rump touched the trigger-plate, and the door clanged down in front of his nose. He was caught!

'For heavens' sake, where's Phil?' Gordon hissed as John and I ran up to him. 'We haven't got much time. Ah, here he is.'

Phil ran up. 'I saw it all from down the lane,' he whispered excitedly. 'But it was lucky I'd stopped to watch the fox that was watching you! There were two sacks, so I've brought both of them.'

'I think they'll be needed,' Gordon replied. 'I don't like the way he's knocking himself about in there.' The badger was reacting strongly to our presence, and had several times run into the end of the trap as though it weren't there.

We arranged one of the sacks over the end of the trap, slid the door up, then tried gently to tip the badger into it, but he clung to the closed end, his wonderfully patterned head raised like a cobra's, ready to take on all comers. Then, as if suddenly deciding that the dark folds were better than the steel mesh around him, he dropped down into the sack. As the sudden extra weight jerked the sack off the framework, both Phil and I, with a single thought, stepped forward to grab the open end – a wasted effort on my part, as I merely cannoned off him and landed flat on my back in the nettles. As I leaped up again, already beginning to smart from a million stings, Phil had hold of the sack, screwing the top round and looking scared stiff.

'Hurry! It's trying to get out. Hold it! Hold it!'

'Here's the slip knot,' John said, handing a mass of rope to Gordon, who quickly secured one end round the heaving sack, which was now dragging Phil in a sort of ballet dance up the lane.

I took a quick look at the nearby bedroom windows. If any early risers caught sight of these mysterious goings-on in the half light, we could expect the Police to arrive before the second badger!

I have described in detail our first hand-to-hand encounter with a badger to show how much can happen in a short time. From the moment when the trap clanged shut to when we attached the captive badger to a tree down the lane, less than two minutes had elapsed, and we were ready to use the same methods on the second animal.

As the nettle patch had been trampled almost out of existence and, worse still, contained the warning, musky scent of frightened badger, we set the trap this time inside the garden where the first badger had come out. If this second animal intended to return at all, it must already be in the vicinity of the wooded gardens and there was only one way to panic it into the ambush: we would have to 'drive' the private property.

The road bent sharply at the end of the third house and, running to this spot, we vaulted the fence and found ourselves in a maze of rhododendron bushes.

'Spread out and cover as much ground as possible . . . What's that?' A deep baying had broken out behind us.

'The Hound of the Baskervilles!' said Phil and increased his pace.

Navigation was difficult in the dense bushes, so it was just as well that the bloodhound or whatever it was didn't seem to be getting any nearer. We continued to fan out, losing sight of one another. I was battling through a section of this luxuriant suburban jungle when Gordon called out, 'There it goes!' followed by, 'Stand still and listen.'

There was no need to strain our ears. A badger travelling in thick undergrowth is unmistakable, the noise being exceeded only by that of human beings. The animal was running in the right direction, but would it take the badger path? If too frightened, it would probably just keep running in thick cover away from us.

Then the jangling, tinny sound of the trap's trigger-plate ended with a clanging thud.

'It's in! We've got it! Go back the way we came, it's quicker.'

The tethered Baskerville hound was now making enough noise to rouse the entire neighbourhood, but this time we executed the trap-to-sack transfer operation smoothly, and within minutes were ready to leave.

We put the most lively sack inside the iron cage, Phil and I taking the front and Gordon the rear. John insisted that he was in full control of the second badger, which was lying passively in its sack over his shoulder and bumping gently against the seat of his trousers.

But his confidence was short-lived. He suddenly arched his back, gave a piercing yell and began high-stepping down the lane! It was not a severe bite, John assured us presently. After all, the badger hadn't been able to take aim, but we noticed that a soft cushion appeared on the driving seat of his car next day, and remained there for a week! He now claims, proudly, that the teeth marks can still be seen.

We reached the car without further incident, put both badgers in the trap, and lashed it back on the roof rack with the seemingly endless clothes-line. As the engine started, we took a last look down the leafy lane towards the rubbish dump that had once been a badger set.

Morning mist was now thick everywhere, but through it we could see a small group of people and a large dog – the Hound? – standing looking at us as we drove away. We wondered if the man who had poured liquid cement down the holes was among them.

I won't go into details of that ride; it was a very anxious one. No wild animals should be confined for a moment longer than necessary, so we brought our charges to the farm as quickly and as gently as possible, then carried the now extremely heavy trap to the strip of woodland containing the empty set.

The set had to be empty, and we checked thoroughly for signs, to be sure of it. Even the easy-going badger will attack a 'stranger' if it is thrust upon it too suddenly.

Everything was all right. We laid the sacks on the mound of the biggest hole, untied them and stood back in a semicircle. Soon they began to move. The big male was first. He pushed his way out quickly, confidently, pointed his handsome striped face at each of us in turn, then at the hole, and paused – nocturnal beauty in a world of dappled sunlight.

It is strange how wild animals never lose their dignity, no matter how humiliating the circumstances.

The sow had discovered the mouth of her sack, and now she ran to him. I like to think that he had been waiting for her, because it was not until this moment that he approached the opening of the set. The broad head faced us briefly for the last time; then, turning, he led his mate down into their new home.

10

SNOWBALL HAS VISITORS

AS THE Iron Samaritan went, for the time being at least, into retirement in one of the farm buildings, and John returned to the exacting job of 'wild life policeman' in his own district, we chose a quiet moment to hold one of our coffee break meetings in the hide, to discuss how things were going in the Snowball area.

Firstly, Snowball himself. He seemed quite happy at the farm, though preferring his own company to that of the boisterous, romping cubs, now almost full-grown. He could usually be seen soon after dark heading for the beech-wood path, leaving the sow to play for a while with the family. She would join him later.

Next on the agenda: during the following week two distinguished naturalists were coming to see the white badger. We must make sure the set received no disturbance on or just before these occasions, or they would be unlucky!

Then there was our little movie film of Snowball. Although taken on eight millimetre film the television people wanted to use it in a programme about badgers. We would have to contact Pete Bowman, who kept the original spool.

Life was becoming a scramble!

Another thing that had come up recently was a request from the local Field Club (we had all joined now) to take some of the junior members in groups to see their first badgers. Bev would have to be asked about this, but of course he wouldn't object.

Last, but by no means least important, was the problem of the rabbit catcher. We had no idea yet who he was, but Bev had reported gunfire and headlights flashing across the fields in the middle of the night, and we had found tyre-tracks and empty cartridge cases in the cow meadow near the set. Gordon had found out from the Police that it is illegal to fire a gun after dark, and they had promised to ask the night-patrol crews to keep a look-out.

Before calling the meeting to a close Gordon demonstrated his latest gadget, a device for determining exactly when badgers arrived at the hide – important to know on nights when none of us could be there. He had drilled a hole in the top of an old clock and pivoted a lever inside so that in one position it rested on the fly-wheel and stopped the works. Now he tied a monkey nut to one end of a piece of cotton and threw it out of the window. Then he gently tightened the cotton and attached its other end to the lever which was well away from the fly-wheel and in the 'Go' position.

'Want to try it out, Gary?'

It was too simple – but it worked. I had hardly lifted the nut from the ground when the lever fell over and stopped the clock.

That night the clock stopped at nine twenty-two. It was extremely useful information, but I would rather have been there to see for myself!

Norah Burke, the first of our important visitors, had been corresponding with Gordon for some time. Author of many fascinating tales of wild life both in Britain and in India, her first home, we knew her to be a tough, efficient field naturalist. It was just as well that she had these qualities for, after driving nearly a hundred miles, she arrived at the farm in the middle of the worst thunderstorm that I can remember. To our delight she seemed totally unconcerned at the appalling conditions and merely remarked that it was 'rather wet'. In the shelter of the egg-room canopy she began to don her 'working' clothes: heavy, neutral-coloured storm-coat, wide-brimmed hat, gloves and gumboots. I hope I didn't stare too rudely at the transformation, but I was already overcome with esteem for this genteel, quietly-spoken lady, and no longer felt incredulous of the reputation she had acquired for woodcraft. It was said that in the forests of East Anglia near her home she regularly approached that shyest of all British animals, roe deer, to within a few *feet*!

In cascading rain we showed her the set from a respectful distance, pointing out the position of holes and most-used runs. She nodded understandingly and, shaking her fist in mock anger at the black clouds, crept to her selected position under the sycamore saplings near the main entrance.

We didn't give much for her chances of seeing *anything*, let alone Snowball, and went unhappily into the hide to wait out the agreed two hours. She could have come in with us, of course, but we had had to admit that, though the cubs usually came into view, Snowball had always shunned our little house with its bright lights, and it was Snowball she had come to see.

During the next hour nothing moved in the length and breadth of the light-beam. Then we all nudged each other together as a bedraggled figure picked its way sedately through the brambles and gave a triumphant 'thumbs up' in our direction as it neared the hide.

A few minutes later, sitting in front of the electric fire and clasping a mug of hot coffee, she told us of her success.

'When I settled in I thought things were just about as hopeless as they could be: only one hole in sight, dense undergrowth hiding everything, and rain still pelting down. I felt like a house with water shooting off in all directions. But then, while it was still daylight, a badger cub appeared at the hole. He came straight up and apparently it didn't worry him in the least that there was a bungalow only thirty yards away! He climbed the bank, then came striding purposefully down it and back into the set. Isn't it amusing how importantly the cubs behave – as if they had a dozen different things to see to? Next, all four cubs came up the bank together. After that the sow appeared. I thought. "Now, if only—" Then – a big white head, and out he came! He's a big badger, isn't he? And I nearly split my head off with a grin of delight. He went round and about the hole for a time, eventually passing below me behind a screen of nettles. He melted along like a white cloud – like the ghost of a badger – then he was gone. When they'd all finally disappeared I came away.'

As it was much too late for Norah to return home that night, and as Gordon's mother already had company staying, she slept at our house. Before leaving next morning she gave each of us a copy of her new book, *The Midnight Forest*, and invited us all to visit her hunting grounds whenever we liked.

Meanwhile, the other two had begun to bring some of the Field Club children to the hide in small groups, and I was able to take my turn at escorting them and 'arranging' for badgers to appear at the right time by putting out a suitably tantalizing trail of nuts. So well did the children observe the rules of silence that not once did we have a blank watch, though the 'clicker' motor came in useful, producing sufficient noise to cover the inevitable whispering that arose when the first badger hove into view. Even Simon, Gordon's six-year-old nephew, who I thought would be sure to make some noise, not only sat like a statue for nearly an hour, but was the first to point excitedly as the old boar lumbered out from the bushes.

There were other diversions in between the badgers' visits, the most amusing being the field mice who scampered into the bright light to snatch up nuts and pieces of bread. One of these, a veritable little Hercules, managed night after night to push, pull and carry a whole, thick slice of bread a distance of several yards into the

darkness – each one enough food to last him a week! But the climax, as far as I was concerned, occurred on the last of the club's visits. A young lady of about thirteen, seated in the corner with her nose pressed to the window, suddenly jerked back and fell off her stool as the wickedly handsome face of an incautious fox materialized round the corner of the hide and peered at her, for the briefest instant, at no more than three *inches* from the other side of the glass! We all felt envious of her good luck.

Soon afterwards we had the other important visitor. Doctor Ernest Neal had had a standing invitation from us since Snowball's early days but, living even farther away than Norah and being busily engaged on scientific work, he had been unable to accept. Now he was staying a few days in London and had phoned to say he was looking forward to the chance of seeing his first albino badger. Ernest Neal was, of course, the Expert. His new naturalist book, *The Badger*, was – and still is – the 'Bible' of all badger watchers. But apart from the thrilling prospect of meeting him, we felt a kind of special responsibility. He had never seen a white badger! Consequently, when he met us at the farm I was a bundle of nerves, poking our tattered copy of *The Badger* at him for his autograph, and forgetting half the questions I'd been saving up to ask. He seemed particularly interested in our 'luxurious' hide, and suggested several animal-behaviour experiments that we could do from it.

'And be sure to keep detailed notes of everything that happens,' he advised. 'Because apart from being scientifically valuable, they'll prove useful if *you* decide to write a book!'

Then, after examining the set and testing the wind, Dr Neal took up, almost identically, the same position that Norah had without a word of advice from us, and settled down to what he later described as a 'superb performance' by Snowball and his family.

'Thank you for a delightful experience,' were his last words as the car moved away down the track. 'Let me know how Snowball gets on and if I can be of any help.'

And in gratitude to Snowball for not letting us down, I crept back and threw four extra handfuls of nuts on the beech-wood path.

A MATTER OF LIFE AND DEATH

WILD WAYS are sometimes difficult for humans to understand. A few days later Snowball and his mate left the farm set to their now capable cubs, and 'moved house' to an old fox-earth a quarter of a mile away. We might not have known for weeks where they had gone but for the lucky chance of Phil's early-morning watch from the observation ladder. These ladders, which we had built from living timber, were positioned permanently at several of the sets and were extremely useful. Averaging ten feet high, with a comfortable seat on top, they enabled us to watch in any wind, our scent blowing always *over* the animals with the same effect as from the hide chimney.

After Phil noted the safe return from the beech-wood of all four cubs and the old boar, he was amazed to see Snowball, followed by the sow, skirt the set and begin to cross the grassland to the north of the farm. Dawn was breaking. It was already late for badgers to be above ground, and Phil could tell by their steady trot that they had no intention of coming back.

Intrigued and a little worried, he descended the ladder and moved to the iron boundary fence just in time to see, through binoculars, the two badgers entering a large thicket of gorse. The angle of slope was such that he could see all round the patch and he didn't see the badgers come out. We knew only too well the two-hole fox-earth it contained. Fox cubs had been born there last spring and our attempts to study them had been frustrated by the impenetrable barrier of thorns. Cutting a way in, even to reach Snowball, was, of course, unthinkable, but this time I found one place where, by tunnelling on my stomach through the maze of stems, I

The observation ladder

could gain entry, and I soon found myself worming into the closest watching position of all time, with my nose pressed against the cast-out sand, four feet from one of the holes! Was it *too* close? *I* would have to be the one to find out, for both Phil and Gordon were too bulky to enter my tunnel.

And so began, for me, the most fascinating period of the Snowball adventure. I don't know why the badgers never scented me at such short range, but a freakish air-current always seemed to blow in my face. Evening after evening I crawled to the holes, witnessing each stage in the formation of a new badger set.

Each watch is remembered by at least one incident or 'highlight', and some of these, as on the first night, were pretty frightening. I had been lying in position about twenty minutes – half on my side and propped up on one elbow – when I felt a repeated thudding below ground. A second or so later the hammering sound suddenly grew in volume, and in a cloud of dust and sand a large fox catapulted from the hole, sailed over my head and crashed out of sight through the bushes. Para-lysed for a moment, I remember blinking bits of twig from my eyes and then looking again, blurredly, at the hole. Snowball stood there spattered with dirt, his nos-trils opening and shutting angrily with each audible breath. I could only guess that the fox had gone in just before I arrived and perhaps even *because* I had arrived – a risky thing to do with badgers in residence, and prob-ably interpreted by Snowball as a 'take-over bid' and dealt with accordingly!

Although vision was restricted from my cramped position, as I could do little more than roll my eyes, I was able to see much of what went on when the animals first emerged, though it was strangely unnerving to look *up* at them. Usually they stood for a minute, nibbling and 'kissing' each other and making little whickering noises. Then one or the other would move off to gather bedding, returning with a great armful of the stuff, usually grass and bracken. Shuffling backwards with a rapid, cater-pillar-like action, it would follow the winding, newly-formed path with accuracy and on reaching the hole would tumble into it, usually losing part of its precious load.

Often, while bedding went in at one hole, sand was being excavated at the other, and when this happened at the one immediately above me I sometimes felt in real danger of being buried alive, a victim of the badgers' industry! The rate at which they can remove earth is astonishing, and so is the method they use, which is prac-

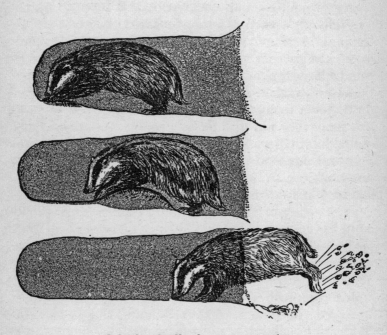

A badger's digging movements

tically indentical with their movements when collecting bedding. Earth is scraped up with the front claws, hauled up backwards in the forearms, and passed to the back feet, which kick it clear. It was always during this final kick that I shut my eyes and hoped there were no large stones amongst the deluge of sand!

The only other incident which nearly caused me to retreat was the sow's discovery of a wasps' nest at the edge of the gorse! I should have known that, as wasps can't find their way in the dark, there was no real danger, but I lay perspiring for many minutes, listening to a chorus of furious buzzing while she noisily munched everything edible, and probably much that wasn't!

I watched this happy pair for nearly a month – seventeen 'sessions' in all – before falling prey to the carelessness which so often follows repeated success. For the past few nights I had been in the habit of 'calling' Snowball back to me when he was out of sight in the bushes by scratching like a badger, a trick which we had found to work quite well if used in moderation. It is done by scratching the short hairs at the back of the head as quickly and as hard as possible, and I had just enough room to do it. Snowball would come padding back immediately to investigate, enabling me to feast my eyes once more on his strange beauty.

Although we have plenty of evidence that badgers' vision is better than many people believe, it was not really surprising that I was never recognized in my camouflaged position, for their eyes seem incapable of focusing in sharp detail. But when I tried to call Snowball for the second time in five minutes, his sharp ears must have located the direction and distance of my scratching with some accuracy. He walked slowly to the top of the mound and gazed straight down into my face, his ruby eyes glinting with the beginnings of awareness. He backed away, sneezed, shook himself, then came forward again. This time, his suspicions confirmed, he spun round and vanished into the hole.

As the cloud of dust from his agitated fur drifted over me, bringing with it that now familiar but still exciting sweet, musky aroma of badger, I knew that I had pressed my luck too hard. The watching must end now, for a while, or they would become nervous and unhappy. The next day I sealed in my tunnel with twigs and branches and left them in peace.

Now that our last batch of visitors had gone and life had returned to a slightly more leisurely pace, we turned our thoughts again to the ever-present threat of the rabbit catcher. To find out the extent of the danger we spent several week-ends investigating every wood, hedge and coppice within a three mile radius of the farm for signs of disturbance, and by the end of December had some very worrying results. Of fourteen rabbit 'buries', six had been blocked up and gassed; ten out of twenty-three fox-holes similarly treated; and, horror of horrors, two occupied badger sets 'killed' in the same way. In each case the same tell-tale evidence was found – empty cyanide cans thrown carelessly into the undergrowth. Although knowing that it is legal to gas rabbits and foxes, which can sometimes be pests, we felt we would like to have a few words with whoever 'mistakenly' poured the lethal powder into a badger set.

We called on Mr. Thomas, who owned the land next to the farm. Could he tell us who the culprit was?

'No, I wish I could.' He sounded quite angry. 'It's probably the same chap who drives round my fields in the middle of the night, but he turns up at such odd times we're not sure how to catch him. My cowman tells me he's heard gunfire on the last two Saturday nights.'

This gave me an idea. 'Could we lie in wait for him in your Dutch barn near the gate? That's where the tyre tracks come in.'

Mr. Thomas laughed. 'Wait out all night in this weather? There was a frost last night; you'd perish. Wait a minute, though.' He paused. 'I suppose you *could* survive in amongst all those straw bales. When do you want to do it? Next Saturday? All right, but if you light a fire, keep it well away from the barn, and if you *do* see our trespasser, don't try any heroics. Remember, he's got a gun. Just get the car's number if you can, and I'll do the rest.'

Poor old Gordon. He didn't look at all happy at first.

'I've slept out once before in winter,' he warned. 'You've no idea how cold it gets. It can actually kill you.'

But after singing, 'Straw bales, straw bales,' at him a few times, we brought him round to our point of view – that it was worth it for the chance of ambushing the greatest danger to badgers that we had yet encountered.

The following Saturday afternoon was grey, dry and cold, and in a gusting north-east wind, which seemed likely at any minute to blow the corrugated-iron roof clean off the rickety, open-sided 'barn', we prepared for the fourteen hours of darkness which lay ahead.

Gordon drove Betsy into the field, parked her on the 'blind' side of the barn and unloaded sleeping bags and cooking equipment, while Phil and I arranged some of the bound straw bales into convenient steps to the top of the stack. Here, under the roof, we built a huge, square nest, with walls two bales high and including, at the downwind end, a narrow opening leading to the steps. It seemed an ideal fortress, snug and wind-proof, yet providing a clear view of the gate and the entire field. Indeed, by standing up with elbows resting on the bales to steady the binoculars, I could see the chicken farm buildings through the trees. As I watched, a pin-point of light began twinkling on the horizon; our hide lamps had come on.

A short distance away from the barn, in the shelter of the roadside hedge, another light burned as Gordon's little cooking stove signalled the approach of our first meal. Soon we were all gathered round it, loading and unloading the sizzling frying-pan and getting in each other's way looking for the plates – which had been left behind! Because of this, the food had to take the form of gigantic sandwiches containing a whole fried egg, a sausage and a rasher of bacon. Unfortunately, as I compressed mine in readiness for the first bite, the fried egg oozed gently out of the far end and fell with a wet flop on to the toe of my gumboot, causing Gordon and Phil to stagger about laughing for so long that their own sandwiches got cold! We finished with coffee brewed in a saucepan, saving our vacuum flasks for emergency rations later in the night, and as it was now quite dark and beginning to freeze, we climbed back in the nest and crawled into our sleeping-bags, overcoats and all.

To help us forget the cold we talked in carefully lowered voices of memories of hot, sunny days; of that glorious afternoon when, crawling to the rim of the old bomb-crater in Wilson's field, we had looked down on a vixen lying on her side suckling her six cubs. As we spoke of it the shrill voice of a vixen came squalling down the wind outside, answered immediately by the yapping bark of a male. This was January, foxes' mating time, and we hoped there would be cubs in the bomb-crater again this year.

We talked of the experiments we had made to enable us to stalk close to foxes and of the time we had tried, once and for all, to beat the problem of human scent by spraying ourselves with 'Essence of Roses'! The funny thing was that it had actually seemed to work – at least, it had with that big dog fox in the beech-wood. He had trotted past within a couple of yards with no reaction at all, except perhaps for a slight wrinkling of the nose, and we *may* have imagined that!

After an hour or so the conversation petered out, and I lay, warm and comfortable now, listening hard each time a car passed by, and wondering if the next one would be the rabbit catcher. It was still only nine o'clock and feeling wide awake I struggled to my feet clutching the sleeping bag in position round my shoulders and peeped over the edge of the nest.

It was snowing! Huge flakes of it fluttering down in the dying wind. I reported the news excitedly to the other two.

Of course, they had to have a look, though there was little to see – yet! Would it settle? As I sank back again, visions of past winters began to filter through my mind, slipping into a series of dreams in which I once again tracked the nocturnal journeyings of badger, fox, rabbit and weasel through the snow.

I awoke in agony. Phil was standing on my hand! His voice came urgently from somewhere above. 'It's snowing harder than ever. It's settled, too, and looks jolly deep. Let's go down and see.'

'Careful!' Gordon's sleepy voice had a note of caution. 'Remember what we're here for. Don't make footprints away from the barn. I'll be down in a minute when I've had some coffee.'

At the bottom of the steps the icy air gripped like a steel clamp, and we stood peering out at the grey blur that had once been a field. But not for long. There is something about snow: it *demands* to be thrown at some- one. So when Phil bent over to look for mouse tracks at the edge of the straw I lobbed a tiny snowball, which caught him on the ear. The slightly larger one he threw back missed me altogether, but at least showed that he was ready to do battle. Just then Gordon stirred in the nest above, and in silent agreement we both poised with our next missiles ready for the new target.

But they were never thrown.

There had been no vehicles for some time, but now the sound of a powerful engine came to us. It was slowing down, stopping – at the gate!

In a panic Phil and I leaped simultaneously up the steps, and met Gordon coming down! For an instant, in the tangle that followed, my feet slipped off the straw, and I hung over the edge of the stack, suspended by my collar in Gordon's grasp.

His voice rose for a moment above that of the roaring engine as he almost threw me into the nest.

'Quick! Hide yourselves. I've got to get his number.'

The rabbit catcher had opened the gate and now, as Gordon leaped off the top step, we hung over the bales trying vainly to see the car through the blinding swirl of snow. But all we could do was guess at events as its engine rose three times to a scream, but remained in the same position. He was stuck! Once more the invisible vehicle shrieked its protest into the night, then with a grinding of gears the noise diminished. He was backing out. Going away!

Before the drone of the defeated enemy faded into the distance, Gordon was back with us.

'That chap doesn't know his Country Code,' he said. 'I even had to shut the gate after him.' He flopped down and reached for his flask.

'The number! Did you get his number?'

'Oh, yes – but I've forgotten it already. – It's all right,

it's all right,' he added hastily. 'Just get a pencil and paper and shine the torch down over that corner. I wrote it in the snow with my foot.'

After the drama of that episode I thought I would never sleep again, but of course we all did, after a while. In my dreams I heard a high, sharp noise like a staccato note on a trumpet, and I jolted instantly awake. My watch showed a little after five o'clock. Had it been part of a dream? No, too *familiar*. As my fuddled brain tried to separate sleep from reality the call came again. That was it, of course! The 'play' note of a badger!

Heavy breathing from the opposite corners of the nest. This would never do.

'Wake up, Gordon, Phil! I've just heard a badger.'

It had stopped snowing, and the field lay before us like a clean sheet, except for one blemish, one black smudge, about a hundred yards out. We grabbed binoculars. It was a badger and coming our way, its short legs half wading, half ploughing through the six-inch deep snow. When about twenty yards from the barn it turned away towards the hedge, each step now clearly audible as the snow crunched softly beneath its weight. As it disappeared against the black outline of the roadside hedge it began to root about noisily, no doubt searching for small hibernating creatures in the hedge bottom. Then, with an eerie chill sweeping over me, I realized that the 'crump, crump', sound of trodden snow hadn't stopped! Something was still approaching us across the completely blank field! Closer and closer, surely almost up to the barn by now. It was going in. Snowball's wedge-shaped head stood out for a second against the floor's blackness, then, as badger-brain reacted to human scent, he was away. And this time we could see him; a faint snow-shadow bounding back into the violet darkness.

In the hard, bright light of morning when next we

woke the world seemed a different place, so much so that, as we packed our things away in the old sidecar, the rare sight of a flock of siskins feeding in a nearby birch caused no more than a flurry of excitement. All seemed to pale before the magic of the past night.

Within hours the temperature had shot up in typically English fashion and by Monday the snow had gone. We were almost glad, in a way, not to be tempted out, for homework had been piling up and called for a few evenings at home. Gordon undertook to put food out at the hide during the following week but he too had writing to do, so we agreed not to meet again till the next Sunday. We had phoned in the car number to Mr. Thomas on that first snowy morning, so on the Sunday afternoon, assembled once more on Betsy, we chugged along to his big house to find out whether he had any news.

. Mr. Thomas seemed pleased to see us. 'Come in and have some tea, lads. Nice piece of work, getting that blighter's number. We won't be having any more trouble from him. Rabbit catcher indeed!'

Two cups of tea later, and after some rather cunning questions from Gordon, we at last obtained all the facts. The car number had been traced to a large estate some miles away, the owner of which ran a rabbit-control society. Any farmer or landowner could join by paying a fee to the society, whose operators would then visit at regular intervals, keeping rabbit-numbers down on his land by legal shooting and gassing. Our rabbit catcher was merely one of these operators who had become careless to the point of ignoring boundaries and gassing every hole he came to! We were relieved when Mr. Thomas told us that the man been severely reprimanded and warned that one more 'mistake' would cost him his job.

As we were leaving Mr. Thomas called out from the door. 'Oh, by the way ... We only found out who this chap was today, you know, and I was present at the ticking-off. I happened to hear him admit that he had just come back from gassing a couple of fox-holes to the north of the poultry farm. Thought you might be interested. The holes are in a patch of gorse, or something. Cheerio!'

We didn't speak; just got on Betsy and drove her as hard as she would go towards the farm. But I couldn't stop those last words ringing again and again in my ears. 'A patch of gorse ... A patch of gorse ... !'

Gordon dismounted slowly at the bungalow, his face suddenly looking much older as he began to trudge in the direction of Snowball's home. We followed him in single file as he had taught us to nearly three long, happy years ago.

Past the farm set, over the iron fence, and across the now-dead grassland. In the fading light, a quarter of a mile distant, a dark blob on the face of the hillside – the gorse patch. Once Gordon stopped for a moment and looked quickly through binoculars, and as we walked on again I fought to convince myself that I hadn't heard him murmur, 'No – Oh, no!' But soon, in the light of our torches, reality was before us. One of the gorse bushes had been hacked down and beyond the great, gaping wound could be seen the two holes, stamped in and blocked.

We backed slowly away as Gordon tried to comfort us.

'There's still a chance. We'll watch for a while. Come over here by the tree and use my binoculars. You never know with Snowball, he's a clever old thing.' But his voice gave him away. He had no more hope than Phil or me. Somehow I felt especially sorry for Phil, who had been the first to set eyes on our 'miniature polar bear', and now crouched, helpless, beside me.

I don't think any of us knew why we waited, or for how many hours. A hunting tawny owl sailed over us many times like a great bat; mice and shrews squeaked around our feet; and once a fox yapped close behind us as it caught our scent. But no-one moved. Nothing seemed to matter any more.

Then, in the absolute silence there came a noise so faint that I felt rather than heard it: the sound of a tiny pebble rolling down a slope. That was all, but I knew Gordon and Phil had heard it for their bodies had tensed. There it was again: several pebbles. The binoculars seemed to weigh many tons as I strained to lift them slowly, and I knew then that the desperate hope that had slept at the back of our minds was now more than a hope: that the walls of the freshly-dug tunnels might have soaked up the poisonous fumes!

The binoculars came into focus on the despoiled earth at the precise moment when, with a great heave, it bulged, broke, and avalanched down into the bushes. And there, poking out of the shattered tunnel, was the

very dirty but unmistakable head of our white badger!

Even now, in the presence of death, he was cautious, testing the night air and looking this way and that. Then he was out, the little sow at his heels, running into the long, dead grass. There they stopped, sneezed several times, and shook the sand from their coats.

The little sow began to move, slowly at first, then breaking into a trot, heading back in a straight line towards her first home, the farm set.

Snowball watched her go, hesitating. Then, looking back once more at his ravaged home, he turned slowly, picked up his mate's scent trail, and ran after her. We watched his pale form growing ever smaller till it was just another patch of moonlight on the flat, dead grass. . . .

POSTSCRIPT

The end of the white badger story? Not by a long way! But after the gassing incident we decided that Snowball had earned a complete rest from human attentions. We would not try to see him again for a whole year.

The promise was kept – though we could not resist the occasional daytime check on his set – noting tufts of white hair on the perimeter fence as a sure sign of his presence. Once, on a late-night walk along the woodland path, we saw him padding towards the fox bank feeding grounds and this was talked about for weeks! When the year was up, the acquaintanceship was renewed and more white badger happenings began to be scribbled into our notebooks.

Snowball always seemed to have the knack of getting into (and out of) trouble. There was the time when he became caught in a snare set for a fox and dragged the heavy tethering stake about for a week before biting his way free. That nearly cost him his life but, as if to show that he bore no ill will towards humanity, he (let us be honest, perhaps unknowingly), saved the life of Steve, Bev Mote's youngest son. The eight-year-old boy had wandered off one autumn afternoon and became hopelessly lost in the big fields to the north of the farm. Night came and with it an early frost. Weak from hunger, cold and struggling through the long grass, Steve lay down and slept – oblivious to the distant calls of his family, the police and us. Late in the night he awoke to the sound of an animal sniffing loudly a few feet away and rousing himself with a great effort, saw the white badger moving purposefully away. On an impulse, Steve crawled after him till, as warmth returned to his frozen limbs, he was able to rise to tip-toe. And thus the strange procession began the long journey back to the farm set where burned the welcoming lights of the bungalow – and safety.

Was Snowball merely making his routine early-morning return, or could the rescue have been deliberate? We shall never know.

When we last saw him, Snowball had reached the ripe old age (for badgers) of ten years. We believe that, like most of his kind, he died peacefully in a comfortable bed, deep underground where other badgers would have buried him by sealing the chamber entrance. But who knows? The area of woods and fields around High Beeches is large – and there are sets which we have had no time, as yet, to watch. If the white badger is still alive we must surely re-discover him – and that would indeed be another story. . . .

Gary
Philip
Gordon

If you would like to receive a newsletter telling you about our new children's books, fill in the coupon with your name and address and send it to:

**Gillian Osband,
Transworld Publishers Ltd,
Century House,
61-63 Uxbridge Road, Ealing,
London, W5 5SA**

Name...

Address ....

....

....

CHILDREN'S NEWSLETTER